The Microeconomics of Product Innovation

The Microeconomics of Product Innovation

by

Paul Stoneman
Emeritus Professor, University of Warwick, Coventry, UK

Eleonora Bartoloni
Italian National Institute of Statistics, Milan, Italy

Maurizio Baussola
Dipartimento di Scienze Economiche e Sociali,
Università Cattolica del Sacro Cuore, Piacenza, Italy

UNIVERSITY PRESS

Great Clarendon Street, Oxford, OX2 6DP,
United Kingdom

Oxford University Press is a department of the University of Oxford.
It furthers the University's objective of excellence in research, scholarship,
and education by publishing worldwide. Oxford is a registered trade mark of
Oxford University Press in the UK and in certain other countries

First Edition published in 2018

Impression: 1

Published in the United States of America by Oxford University Press
198 Madison Avenue, New York, NY 10016, United States of America

British Library Cataloguing in Publication Data
Data available

Library of Congress Control Number: 2017949306

ISBN 978–0–19–881667–6 (hbk.)
978–0–19–881668–3 (pbk.)

Printed and bound by
CPI Group (UK) Ltd, Croydon, CR0 4YY

FOREWORD

By
Paul Stoneman

Over the past thirty or so years I have written (or edited) a number of books on the economics of innovation and technological change, starting with Stoneman (1983) and most recently publishing Stoneman (2011). None of these books is either a simple textbook or a dedicated research monograph; rather each one sits between these two extremes, in an attempt to provide an insight into the relevant issues. Their contents have been aimed at several audiences, with different parts aimed at (i) stretching a good third-year undergraduate, (ii) providing a foundation for policy professionals from various disciplines who work in government, and (iii) offering a rigorous foundation for both postgraduate students and fellow researchers in economics. All throughout, the view has been that, to a large degree, economics has not given sufficient attention to the microeconomic analysis of innovation and technological change. These texts represent one attempt to stimulate such interest further.

Stoneman (1983) was one of the first attempts to bring together the disparate analyses in the growing field of the economics of technological change. The emphasis in that book on the positive aspects of the subject was supplemented by Stoneman (1987), where the welfare and policy side were addressed. Stoneman (1995), which I edited, was an attempt to update this material (which reflects how the literature had since expanded) by collecting a number of contributions commissioned from distinguished authors in the area who could complete the task more effectively or more efficiently than I alone could have. Stoneman (2002) was an attempt to provide an overview of how economics has contributed to the analysis of the diffusion of new technology and, in particular, how my own published research in this area (in which I have specialized) fits into the field. Stoneman (2011) had a much more evangelical tone than previous books in that it was an attempt to raise the profile of "innovation in form" alongside that of "innovation in function," largely (but not exclusively) by considering innovation in the creative industries, which economics had very much neglected. It was while reflecting upon that book that it occurred to me that it had an additional message: most of the economic analyses of innovation that I had been studying and writing about for many years concerned the development and introduction of new processes, whereas I had just completed a work that was concerned more with product innovation, that is, with the development and introduction of new products; and, to many commentators, especially in government, it was this aspect that mostly distinguished innovative activity. The more I thought about this the more disappointed I became with what I could access in the literature on the economics of product innovation. It was time to return to the word processor and to produce this, my final book. Again, my intentions are the same, namely to offer an insight into the approach of economics to various aspects of innovation and technological change and thereby to stimulate such analysis. In this I have been very ably assisted by Eleonora Bartoloni and Maurizio Baussola, two of my former doctoral students, who answered my call for assistance in completing this work. Maurizio has gone on to a very successful academic career of his own, researching the economics of innovation. Eleonora is senior researcher at the Italian National Institute of Statistics (ISTAT), where she is in charge of R&D and other business statistics. She is also a lecturer in economics.

Our audiences for this book are slightly different from those targeted in the past. In this volume attempts have consciously been made to keep the mathematical content to an absolute minimum. This is partly in order to make the volume more accessible to business and management students (and not just to economics students), in particular MBA students; besides, as Stephen Hawking has observed (Mlodinow and Hawking, 2008), the use of mathematics is a sure way to lose an audience, which we do not wish (or wish to perhaps continue) to do.

CONTENTS

LIST OF TABLES xi

1 Introduction 1

2 Some Initial Definitions 8

2.1 What is a product? 8
2.2 What is product innovation? 11
2.3 Industries and product innovations 14
2.4 Conclusions 15

3 Sources of New Products 16

3.1 Introduction 16
3.2 Research and development 17
3.3 Design activities 19
3.4 Creativity 20
3.5 Knowledge flows 21
3.6 Imports and FDI 24
3.7 Emulation 24
3.8 Conclusions 30

4 The Extent of Product Innovation 32

4.1 Introduction 32
4.2 Innovation surveys 32
4.2.1 Europe 32
4.2.2 The United States 41
4.2.3 BRIC and other countries 43
4.3 Product launch data 44
4.3.1 Introduction 44
4.3.2 Pharmaceuticals 44
4.3.3 Laundry detergents 46
4.3.4 The food industry 47
4.3.5 The film industry 48
4.3.6 Book publishing 51
4.3.7 Recorded music 53
4.4 Scanner data 54
4.5 Industry maturity and product innovation 57
4.6 Conclusions 59

5 The Demand for a New Product 62
5.1 Introduction 62
5.2 The base case: an original, producer durable, monopoly supplier 63
5.3 Consumer durables and non-durables 67
5.4 Innovations new to the market 69
5.5 Conclusions 71

6 Capacity Creation, Pricing, and the Promotion of Product Innovations 73
6.1 Introduction 73
6.2 The monopoly supply of an original product 74
6.2.1 Pricing 74
6.2.2 Promotional activity 76
6.2.3 Capacity creation 76
6.2.4 The location of production capacity 78
6.3 New-to-market product innovation 80
6.3.1 The monopolist as new-to-market innovator, horizontal product innovation 80
6.3.2 The monopolist as new-to-market innovator, vertical product innovation 82
6.3.3 The monopolist as new-to-market innovator, product promotion and capacity creation 83
6.3.4 The monopolist as new-to-firm innovator 84
6.3.5 Many new-to-market innovators: horizontal product differentiation 84
6.3.6 Many new-to-market innovators: vertical product differentiation 85
6.3.7 Many new-to-market innovators: product promotion and capacity creation 86
6.3.8 New-to-firm (but not new-to-market) innovation 86
6.4 Conclusions 87

7 The Incentives to, and Constraints upon, Product Innovation 88
7.1 Introduction 88
7.2 Why undertake product innovation? 89
7.3 How much innovation? 91
7.3.1 A general approach 92
7.3.2 Horizontal product innovation 94
7.3.3 Vertical product innovation 99
7.4 Risk, uncertainty, and timing 101
7.5 Constraints to innovation 103
7.5.1 Costs of innovation 104
7.5.2 The availability of qualified labour 104
7.5.3 Risk and finance 105
7.6 Conclusions 107

8 Empirical Evidence on the Determination of the Extent of Product Innovation 108

8.1 Introduction 108
8.2 Survey data 109
8.3 The determinants of innovativeness 112
8.4 Constraints on innovation 117
8.5 Entry, start-ups, foreign direct investment, and imports 119
8.6 Conclusions 122

9 The Diffusion of Product Innovations 124

9.1 Introduction 124
9.2 The global perspective: the extensive margin 126
9.3 The intensive margin 128
9.4 Intra-firm and intra-household diffusion 131
9.5 The diffusion of the production of product innovations 132
9.6 Conclusions 134

10 Product Innovation and Firm Performance 136

10.1 Introduction 136
10.2 Measuring firm performance 138
10.2.1 Accounting-based measures 138
10.2.2 Market-based measures 140
10.2.3 Cost accounting measures 140
10.2.4 Other financial measures 141
10.2.5 Non-financial (subjective) measures 142
10.3 Some estimates of the impact of product innovation on firm performance 142
10.3.1 Product innovation and profitability 142
10.3.2 Product innovation and the market value of firms 146
10.3.3 Productivity, product innovation, and R&D 149
10.3.4 Product innovation and exporting 152
10.4 Innovation complementarities 156
10.5 Product portfolios, firm performance, and the product life cycle 159
10.6 Product differentiation and firm performance 161
10.7 Market orientation and the product development process 163
10.8 Conclusions 165

11 Product Innovation and Price Measurement 168

11.1 Introduction 168
11.2 Patterns of product creation and destruction 168

11.3 Product innovation and price indexes 170
11.4 Hedonic methods 174
11.5 Conclusions 178

12 Product Innovation and Welfare 180
12.1 Introduction 180
12.2 Product innovation and consumer and producer surplus 180
12.2.1 Repeat purchase products 180
12.2.2 Durable products 185
12.2.3 The distribution between consumer and producer surplus 189
12.3 Product variety 191
12.4 Conclusions 193

13 Product Innovation and the Policy Dimension 194
13.1 Introduction 194
13.2 Intellectual property rights and protection 196
13.3 Tax incentives to product innovation 203
13.4 Government grant-based policies 207
13.5 Public contracts and government acquisition 208
13.6 Addressing financial constraints 211
13.7 Access to knowledge 213
13.8 Infrastructure 213
13.9 Regulation 214
13.10 Foreign direct investment 215
13.11 Subsidies to the demand for product innovations 217
13.12 Prizes 218
13.13 Conclusions 219

REFERENCES 221
INDEX 241

LIST OF TABLES

3.1	Sources of information for innovation	23
3.2	Leading applicants for design registration, OHIM, 2011	28
3.3	R&D spending and IP applications worldwide, 2002–11	30
4.1	The percentage of firms that have undertaken product innovation, 2012 CIS surveys	34
4.2	The percentage of sample firms innovating in goods and services, 2010 CIS 7	35
4.3	Proportion of sample firms undertaking product innovation (%), CIS, 2004–10	36
4.4	Product-innovating firms: goods new to firm and new to market, 2010 CIS 7	37
4.5	Product innovation in the United Kingdom, 2009–15 innovation surveys	38
4.6	UK enterprises engaging in product innovation, 2012–14, and shares of total turnover arising from product innovation, 2014	40
4.7	Companies in the United States reporting innovation activities, by industry, 2006–8 (%)	42
4.8	Product innovation, manufacturing firms, various countries, 2011, % of sample	44
4.9	Summary of number of new drug applications (United States, 1990–2011)	45
4.10	FDA new drug approvals by decade	46
4.11	Total initial trademark registrations, NICE class 29	48
4.12	Trademark registrations and renewals, all classes, 1970–2009	49
4.13	World production of feature films, number of films made, 2005–9	50
4.14	New books published per country per year	52
8.1	Percentage of broader innovators, self-reported perception of potential barriers to innovation, UK Innovation Survey 2015	118
8.2	Obstacles to innovation considered highly important, percentage of respondents (innovators and non innovators, seventeen EU respondent countries), 2012	119

1 Introduction

The prime objective of this book is to consider how the use of economic analysis can guide and inform our search for insight into the generation and adoption of new products, an activity labelled 'product innovation'. Product innovation has received less attention in economics than its cousin, process innovation, a case also made by Lundvall and Vinding (2004), who go on to argue that moving the focus of analysis to product innovation yields different conclusions in various areas. In our view, there is currently no obvious source to which the reader can turn for material on the economics of product innovation, and thus we are filling a gap in the literature. The book is restricted to the microeconomic side of product innovation, for we do not wish to get involved with the main area of macroeconomics that deals with innovation, namely growth theory (see Acemoglu, 2009). This is because growth theory has little interaction with the microeconomics of product innovation, although the two have overlapping interests. We thus leave to others the analysis of the relationship between product innovation and aggregate output growth, employment, wage rates, and inflation—*inter alia*.

The book has several objectives at different levels. The intended audiences are varied, comprising undergraduates, graduates, professional (government and policy) economists, researchers, and teachers. An attempt has been made to keep the mathematical content of the text as limited as possible in order to encourage a wider readership, especially one that encompasses management and business (e.g. MBA) courses, and not only economics students.

For these various audiences, our prime aim is to provide an overview of what we already know about the different aspects of the topic of product innovation. Thus there is a big element of literature review in the text (some of it encompassing the authors' own prior work). This element contains both empirical observations—for example, on the extent of product innovation—and presentations of analytical and theoretical approaches that attempt to model such innovative activity. However, the literature is in fact a rather disparate body of knowledge and thus the authors make attempts to draw together its different strands in order to explore any agreement on important issues that enables conclusions not previously drawn. To a limited extent, the authors also attempt to bring to the topic new insights, both analytical and empirical.

The volume has another objective as well. It is the authors' opinion that, to a large extent, the study of innovation has not received from economics (especially at the microeconomic level) the attention it deserves. Although there has been a very considerable increase in the study of innovation over the past thirty years, the economics of innovation is still not a particularly popular optional course for undergraduate and postgraduate study or doctoral research. As a result, the number of active researchers in the field is disappointing. This book is an attempt to show students and potential researchers that this is a most exciting, active, and rewarding area of economic study and that involvement in it could earn the student a significant return. In this way we may hope that the study of the economics of innovation would thrive in the future.

Not all economists are in academia. There is a large body of economists involved in local, national, and transnational policymaking, data collection, and various types of market analysis. (There are also non-economists in these fields who we hope may find the volume equally useful.) It seems to us that for this audience, and especially for their employers, innovation usually means product innovation, whereas much of the existing literature on the topic is almost exclusively concerned with process innovation. Thus we consider that this book has much of interest for this audience, because it explicitly addresses product innovation and, in doing so, deals with important aspects concerning definitions, data availability, measurement, and policy rationales that cannot be found easily elsewhere.

In order to overview the topics addressed here we will go through the layout of the book and indicate the content of its chapters and how they interact. Chapter 2, in a traditional way, begins with some definitional issues in order to clarify the definition of product innovation. Starting from a discussion of what is a product, how goods and services differ, and what markets are served by different products, the chapter proceeds to discuss what is meant by product innovation, in particular whether newness is a sufficient characteristic to represent an innovation, and, if so, how newness should be defined and measured. Global, local, and new to the firm are distinguished as different types of innovation. Two different types of innovation that are often treated separately throughout the book are identified and discussed here: vertical innovation, where the new product is accepted by all purchasers as better than the existing products (or at least as offering a different quality-to-price ratio); and horizontal product innovation, where not all potential purchasers accept that the new is better than the existing products, but some do. These definitions are used to illustrate how product innovation may not necessarily imply improved or different functionality but can involve changes in form rather than function (Stoneman, 2011). The same definitions are also used to clarify the matter of when a product innovation is economically significant; and a definition based upon market impact is suggested. The relationship between product innovation and (changes in) industry boundaries is also briefly discussed.

Given the basic definitions, in chapter 3 we address the sources of new products, that is, how new products are generated. Much literature on innovation tends to talk almost exclusively of research and development (R&D), as if this were the only source of innovation. It has, however, become more common of late to talk also of design activities not included in R&D as another source, and beneficially so. We discuss both R&D and non-R&D activities as sources of product innovations. We also add creativity, either alone or in conjunction with design and R&D, as a further source, especially for innovations in the creative sector—say, films, books, or music—which is not usually found on the list of sources. To provide further richness, we also present some material concerning the sources of knowledge that may underlie R&D, design, and creativity. We also talk of imports as a source of product innovation in an economy; this category contains both new final products that are being imported and new products that are being manufactured in the domestic economy through overseas direct investment, either by overseas innovators or under license from them. Finally, we talk of emulation. Given that we have allowed innovation to be not only global but also local or new to the firm, many new products will be copies of, or attempts to emulate, new products offered by other suppliers. The discussion of emulation leads us into the topic of intellectual property rights (IPR), which will emerge for the first time at this point. It will help us to build a fuller picture of emulation activities, and will also provide a basis for the discussion of the use of IPR statistics as measures of innovative activity.

Building upon what has been said about the sources of innovation, chapter 4 looks at various indicators of the extent of product innovations. These indicators include data collected by various innovation surveys in Europe in general (the Community Innovation Survey, CIS) and in the United Kingdom in particular; in the United States; and Brazil, Russia, India, and China—that is, in the BRIC countries. Product innovation is shown to be extensive but different across countries, sectors, and firms of different sizes; some products are new to the market and some are new to the firm, the latter being more frequent. Although the survey data are informative, they only indicate whether or not firms undertake product innovation and not the extent to which they innovated. We thus proceed to look at the extent of new product launches in a series of different industries, using other data. This allows for some product innovation to fall outside the Organization for Economic Co-operation and Development (OECD) definitions, on which the survey data are based; and total product innovation may then be better reflected, particularly in non-manufacturing industries. We observe that, although many new products are launched, a high proportion of them are not significantly different from existing products, involving just new packaging or brand extension rather than significant changes in characteristics. We show that different countries may have their own products and product innovations that suit the local culture, although there may well still be much exporting and importing of new products. We note that the rate of failure of product innovation may also be very high, as very few new products succeed and survive.

Finally, industry life cycles to which product innovation activity may be linked are illustrated on the basis that, as has been argued in the literature, industries move through phases on a maturity continuum and it is in the early phase that product innovation is most prevalent. Although there is supporting evidence for this idea, there is no standard length of time for the first phase and the same model may not apply to all industries. It may also be argued that firm entry and exit proceed in phases and in the early phase of maximum entry one may expect more product innovation. Once again, however, the model does not apply to all industries and phases may have different durations in different industries. Nor is it possible to say that industry life cycles are necessarily causal of product innovation—they may just be descriptive.

Chapters 5, 6, and 7 provide an analytical heart to the material. Together, these chapters attempt to analyse (i) the demand for newly launched products, both at a point in time and over time; (ii) supply-side decisions related to capacity creation, the quantity of a newly launched product to be supplied to the market, and its pricing and promotion, all in the light of the revealed demand patterns; and (iii) decision-making re the generation, development, and launch of new products, taking into account these analyses of the demand for and supply of newly launched products. We look at various scenarios related to original, new-to-the market, and new-to-the-firm product innovations; durable and non-durable products; monopoly and oligopoly; and horizontal and vertical innovation.

In chapter 5 we discuss the demand side with close parallels to the various streams in the literature on the diffusion of innovations—for example modelling that encompasses learning effects, buyer heterogeneity, price changes over time, early mover advantages, and network effects. Producer goods and consumer goods are discussed separately, as are durable and non-durable products. Differences between completely new (i.e. original), new-to-the-market (i.e. not original but locally innovative and horizontally or vertically differentiated from existing products), and new-to-the-firm innovations (i.e. not differentiated from existing products, but coming from a different supplier) are also discussed separately. It is argued that, for a completely new (original) product innovation that is durable, demand will be spread over time as a result of learning and risk avoidance, but intertemporal pricing, early mover advantages, the buyers' changing price expectations, and changes in performance will also cause buyers to have different preferred adoption dates, such that demand will reinforce this intertemporal acquisition. It is then argued that, as emulation occurs (i.e. there is further new to-the-market innovation), the wider horizontal and/or vertical product diversity that results will cause demand to increase and ownership to expand. Further new-to-the-firm innovation will reinforce this.

In chapter 6 we explore the decisions of the suppliers to the market of goods that embody product innovations. We discuss suppliers' decisions concerning the pricing and the promotion of new products, and also the creation of capacity to manufacture such products. We also address decisions about the location of production capacity, especially home versus overseas production. Most of our discussion is based on the premise that suppliers are profit seekers and perhaps profit maximizers. Their decisions are thus driven by the profit motive. Decisions on pricing, promotion, and capacity are interdependent, for example pricing low to generate high demand is only rational if there is capacity to supply the market. Thus, although for expository reasons we discuss each topic separately to some degree, the interactions must not be forgotten. We consider both the case where there is a monopoly supplier of an original product and markets where other new products exist, which are both new to the market (but not original) and new to the firm (but not new to market). The thrust of the findings is that firms may well have incentives to further innovate after the launch of the original product, although there are limits to the process that depend on fixed and variable costs and on the elasticity of demand. Similarly, there may be incentives to create further (or less) capacity over the lifetime of a product. Further innovation, either by the original suppliers or by others, tends to lead to lower prices and higher sales and output. This may also imply greater capacity creation (in some locations). We find, however, that, for the originator and for potential new entrants, the incentives may differ according to the internalization and externalization of the impacts of their behaviour. This will tend to imply lower prices when there are many suppliers than when there are fewer suppliers. This will mean that the quantity supplied will be greater if there are more innovators. We have not committed ourselves where the pattern of promotional expenditures is at stake.

In chapter 7—the third chapter in the analytic section—we look at the factors that affect a firm's decision to undertake product innovation. There is an underlying assumption that firms are profit seeking, profits being equal to the difference between the (discounted) net revenues earned from the supply of a new product and the costs of developing and placing that new product on the market. There are many ways in which product innovation and profitability interact, and also a number of triggers that might initiate product innovation in the search for profit. There is also a risk element to innovation that has to be taken into account in the innovation decision. In the context of the issues addressed in chapters 5 and 6 (especially the distinctions made there between original, new-to-market, and new-to-firm innovations, durable and non durable, and horizontal and vertical), we address the general question of why firms or entrepreneurs might undertake product innovation; then we move to consider the extent of the innovation that firms might undertake; and we analyse the type or direction of innovation that can be expected and the timing of innovation activity. Finally, we address issues related to the risks of product innovations and potential constraints to innovative activity; here we consider, *inter alia*, the financing of product innovation.

After the three analytical chapters, the book returns to questions of *what* rather than *why*. In chapter 8 we consider the empirical literature on the determinants of product innovation as a follow-up to the analytics in chapter 7. The theme running throughout the chapter is that firms that are innovative in one dimension tend to be innovative in other directions too. Hence, if one is to explain product innovation, one needs to seek out the characteristics of the innovative firm. This is pursued through survey data and with considerable reliance upon the existing literature.

The aim of chapter 9 is to consider empirical evidence on patterns in the post-launch development of ownership and sales of newly introduced products. This we label the diffusion of product innovations. In particular, the intent is to emphasize the what rather than the why, in other words to reveal patterns rather than to concentrate upon explaining the determinants of those patterns. Three levels of diffusion are identified: the spreading of first use across countries (the extensive margin); the spreading of first use across users within countries (the intensive margin or inter-firm/household diffusion); and the increasing intensity of use by adopters (intra-firm/household diffusion). The details encompass both producer and consumer goods, durables and non durable products, and even management practices. The principal findings are that diffusion takes time, often a considerable period at all levels, and differs considerably across technologies and countries. It is also observed that the diffusion of the production of product innovations may eventually mean that production in countries where early producers were located is eventually replaced by production in countries that were late in the adoption process.

In the three following chapters, 10, 11, and 12, we address the impact of product innovation, considering respectively firm performance, prices, and welfare. Chapter 10 tackles economic literatures that look at how product innovation may impact upon firm performance, and also some managerial literatures that address how managerial behaviour may affect that impact. Various measures of firm performance are discussed, the larger majority being concerned with profitability; and a number of estimates of the impact of product innovation on profitability and market value are discussed. Literatures related to exporting behaviour and productivity are also covered. Issues of innovation complementarities are also addressed.

Chapter 11 explores issues concerning product innovation and prices, with an emphasis upon the impact of innovation upon prices and upon taking account of product innovation in the measurement of price change (both for single classes of goods and for baskets of goods). The main issue to be addressed here is how quality-adjusted price indexes that reflect only inflation are to be calculated when product innovation causes old goods to be replaced by new goods that are often cheaper and of higher quality. Different methods and procedures are discussed and some estimates are provided of the potential biases in measurement that might arise from a failure to adjust in terms of quality. It is also noted that a good estimate of these biases in consumer price indexes is also an estimate of the contribution of product innovation to increases in real consumer well-being.

In chapter 12 we address the issue of how product innovation impacts upon welfare. This is done via two approaches. The first explores the impact of product innovation upon the commonly used measure of economic welfare, calculated as the sum of consumer and producer surplus. Some estimates of this impact are also provided. Next, the topic leads us to discuss whether the maximization of this sum identifies the point of welfare optimality. We argue that, as the split between producer and consumer surpluses is also important, the answer is no. The second approach is to explore how product innovation affects economic welfare via its impact on the degree of product variety in the economy. Throughout the chapter we attempt to address two main issues. One is the impact of product innovation upon welfare; the other is the question whether a free market economy not only will, unaided, produce improvements in welfare but has the necessary incentives to generate welfare optimal outcomes. The results of this discussion will provide some foundation for the subsequent chapter, which deals with policy intervention.

In chapter 13 we look at the policy dimension. We first ask why the government should intervene in the process of product innovation. The three main reasons often used to justify intervention (although not always correctly) are: (i) innovation is good, so more must be better; (ii) there is market failure in the process of product innovation and, left unaided, the market would not generate the welfare optimum (this is based on material discussed in chapter 11); and (iii) policymakers are not satisfied by the rate of product innovation that is being generated without intervention (often using international comparisons to support the view) and believe that their intervention will improve matters (the poor performance not necessarily being the result of market failure). There may be other reasons for intervention, for example health and safety, or the fact that the government is the major customer; but the three reasons listed above are those mainly provided. We then consider the instruments available to the government and the literature concerning whether government objectives will be met. Thus, after an initial discussion of intellectual property rights institutions (patents, design rights, copyrights, and trade marks), we consider other policies, which encompass both supply- and demand-oriented initiatives—for example R&D tax incentives, investment incentives, grant giving, incentives to inward foreign investment, public procurement, buyer incentives, prizes, and regulations.

2 Some Initial Definitions

2.1 What is a product?

It is useful to think of an economy as involving firms (or producers), industries, users (or consumers), and markets. Products are produced by firms, and firms may produce one or many products. Industries are collections of firms that are linked by similar production activities. Markets involve the supply of and demand for similar products.

A product, as the output from a production activity, may be classified as either a *good* or a *service*. The distinction between a good and a service varies according to different authors. There is a line of argument that distinguishes goods and services according to whether a product is tangible, that is, can be touched, or intangible, for example a piece of information. Alternatively, a definition can be made according to whether a product is storable. Services are not storable, whereas goods are. It may be argued that a good can offer a flow of services over time, for example a CD can be listened to many times but no individual listening can be stored. A product may be simple (a nail) or complex (a jumbo jet). It may be bespoke or a one-off, for example a work of art; produced in small numbers for example an expensive sports car; or produced in large numbers, for example a tin of baked beans.

There are standard methods according to which business establishments supplying products may be classified by the type of economic activity in which they are engaged; these methods provide a convenient way of classifying industrial activities into a common structure or set of industries (see Prosser, 2009). Thus the UK standard industrial classification (SIC) of 2007 is divided into twenty-one sections,[1] each denoted by a single letter from A to U;

[1] The UK SIC of 2007 (see Prosser, 2009)—which, it is worth noting, is consistent with both the Eurostat NACE Rev.2 classification and the UN's International Standard Industrial Classification (ISIC4)—defines the following sections:

A: Agriculture, forestry and fishing. **B**: Mining and quarrying. **C**: Manufacturing. **DE**: Electricity, gas, steam and air conditioning supply. Water supply; sewerage, waste management and remediation activities. **F**:Construction. **G**: Wholesale and retail trade; repair of motor vehicles and motor cycles. **I**: Accommodation and food service activities. **HJ**: Transport and storage. Information and communication. **K**: Financial and insurance Activities. **L M N**: Real estate activities. Professional, scientific and technical activities. Administrative and support service activities. **O**: Public administration and defence; compulsory social security. **P**: Education. **Q**: Human health and social work activities. **RS**: Arts, entertainment and recreation. Other service activities. **T**: Activities of households as employers; undifferentiated goods and services-producing activities of households for own use. **U**: Activities of extraterritorial organisations and bodies.

the letters of the sections can be uniquely defined by divisions (denoted by two digits); the divisions are then broken down into groups (three digits); the groups into classes (four digits), and, in several cases, classes into subclasses (five digits). Here is an example: Section C, Manufacturing; division 13, Manufacture of textiles; group 13.9, Manufacture of other textiles; class 13.93, Manufacture of carpets and rugs; subclass 13.93/1, Manufacture of woven or tufted carpets and rugs. In total there are 21 sections, 88 divisions, 272 groups, 615 classes, and 191 subclasses encompassing the production of all goods and services. An example of the production of goods is the economic activity classified as class 23.31: the manufacture of ceramic tiles and flags. An activity producing services is class 68.20: the renting and operating of own or leased real estate.

Markets are the economic institutions where products are offered for sale by potential suppliers and potentially demanded by buyers and where trading may take place. Products offered on markets may be final goods or services (the buyer is the final user), intermediate products (the product is an input into the production process of another product, or even, in some cases, into its own production), or a raw material. Products may fall into one or more of these categories. Buyers may be firms, households, or public bodies. It is common usage to define a product supplied primarily to markets of producers as a producer product and a product supplied primarily to markets of consumers (including public bodies) as a consumer product. Suppliers may be producers, wholesalers, or retailers, a common pattern being that producers sell to wholesalers who sell to retailers who sell on to final users. Suppliers may be domestic or based overseas (i.e. the product is imported), as may demanders, with overseas custom being met by exports (if not by overseas production).

Market classification will not, however, be the same as classifications of production activity. Although we do not have a classification of markets that is the equivalent of the Standard Industrial Classification (SIC), it is obvious that the two will differ. For example, the markets for small commercial vehicles and passenger vehicles will encompass different buyers. However, the manufacture of the two kinds of vehicle is classified in a similar way; both are similar industrial or production activities. In another dimension, the market for certain products may be local, for example regional, national, or international, while production may be otherwise. It is worth noting that, in the absence of a classification of markets, it is most common in data collection to organize data according to the standard industrial classification of production activities. This may, however, provide a misleading picture.

In some literatures products are viewed as collections of more basic performance or other characteristics. This 'hedonic' approach would consider cars, for example, as packages of characteristics encompassing number of seats, fuel economy, top speed and acceleration, type of upholstery, and so on. Mobile phones may be considered in terms of geographic coverage, screen size and clarity, camera quality, memory size, and ease of use. Characteristics need not, however, be functional—appearance, taste, smell, or sound may also be characteristics, for example the smells of different perfumes or the noise of different car exhausts. Some characteristics may be valued positively by buyers and some negatively. Different buyers may also value different characteristics differently. The cost of producing a product would also depend upon the characteristics incorporated in that product; for example, leather seats are more expensive to provide in a car than cloth seats.

In most markets, most products, although of a similar class, will have different characteristics. Such products are classified as differentiated products (although in some markets, labelled as commodity markets, the differences are small). Thus, for example, the market for takeaway meals will encompass pizzas, fried chicken, hamburgers, Indian cuisine, Chinese cuisine, and so on, all of which are takeaway foods but are much differentiated; the market for clothes-washing products will encompass liquids, powders, concentrates and non-concentrates, biological and non-biological products, and so on—again, all directed at the same task but working in different ways.

The literature on product differentiation classifies products into two main categories; in other words it uses two main types of differentiation. In the simple case of two products, when, at a common price, there is no consensus on preference ranking across potential consumers, some preferring one product and some preferring the other, the two products are considered to be horizontally differentiated (Hotelling, 1929). When, again at a common price, there is a ranking, in that all buyers prefer the same product, the products are considered to be vertically differentiated (Gabszewicz and Thisse, 1979, Shaked and Sutton, 1982). For example, a Jaguar car and a Hyundai car may be vertically differentiated, whereas certain Samsung and Apple phones may be horizontally differentiated. As a further example, holidays in Turkey and Spain may be horizontally differentiated, whereas performances of *Hamlet* by the Royal Shakespeare Company and by the local amateur dramatic society will be vertically differentiated.

Differentiated products are frequently branded goods. Economics as a subject has paid much less attention to branding than has, for example, marketing (see Baines, 2014). In much of economics the brand is little more than an identifier that enables one to separate a product from one company from the product of another. In marketing, however, the brand itself has value as an identifier of product source that carries connotations with respect to product quality and performance, derived from association with previous use or purchase that will influence the demand for that product. Brands may be retailer-based (Marks and Spencers), producer-based (Mini), range-based (Tesco's Value range), or even wholesaler-based (Booker). A firm may have a number of brands (e.g. Unilever brands include Dove, Flora, Omo, Lux,

Knorr, Surf, and Sunsilk) and may use the same brand on different products (as in the case of a supermarket's own brands; or, in the example of Unilever brands, Dove is used for both soap and shampoo). The brand of a product may be considered one of the characteristics of that product. The brand attachment may also contribute to the determination of whether two goods are vertically or horizontally differentiated.

2.2 What is product innovation?

Most commonly, product innovation is defined as the supplier-based activity of a firm that offers a new product to the market. The product innovation may be new to the world (global level), new to a market (local level), or new to an individual supplier (firm level). A producer will make a global product innovation when it makes a new good or service available to the world market, in whole or in part,[2] for the very first time. For example, at different times, mobile phones, steamships, computers, and bicycles have all been new goods representing global product innovations; insurance products, supermarkets, hire purchase, and credit cards have all been new services representing global product innovations; industrial robots, payroll software, and electric motors have been global producer product innovations (and as such are often considered to be the embodiment of process innovations); while package holidays, vacuum cleaners, and satellite navigation systems have all been global consumer product innovations.

A local—as opposed to a global—product innovation may be said to occur when a product, although available in other parts of the world, is made available for the first time to a market more limited than the global market—for example the European market, a national market, or a regional market. The local product innovator need not be the global innovator; other suppliers may have followed a lead set by the global innovator. Nor need the local product innovation be produced in the same geographic area as the market that is supplied with it for the first time—the new product may be imported.

Firm-level product innovation occurs when a firm supplies to its current market(s) a product that it has not supplied previously, even though other firms may already be supplying a similar product to these or other markets.

[2] Although in principle a global product innovation in this sense could involve the supply to a different world market of a product previously supplied to another world market, for example personal computers being supplied to the household or residential market when computers had been previously supplied only to commercial users.

Once again, this need not necessarily involve production in the local area (although this is often implied): the new product may be imported from elsewhere or from outside the market area.

One may well expect an intertemporal pattern to exist whereby first, global innovation occurs when a new product is launched by one supplier on one or a few national or regional markets. Then local product innovation will occur through exports from the original supplier, through the establishment of local production facilities by that supplier, or through potential suppliers that imitate the initiator, undertaking firm-level product innovation and establishing production capabilities for the new product with which to supply various new markets or to compete with the global innovator on existing markets.

With respect to their forebears (if such existed), product innovations are sometimes separated into two classes: new and original. Although an original product may be new, a new product may not be original. It is difficult to define what is meant by a completely new product. Bresnahan and Gordon (1997) argue that one approach to the definition of originality is to consider whether the innovation provides fundamentally different value to a user. For example, the first antibiotic might be viewed as a fundamentally new or original good, since the characteristic of having bacteria-killing capabilities did not exist in previous drugs. Similarly, the invention of the phonograph enabled (1) the recording and (2) the replaying of music and of the human voice, neither of which was previously possible. On the other hand, the electronic calculator might be viewed as not fundamentally new, because it offered—although more quickly, easily, and cheaply—the same services that were previously embodied in slide rules, and is thus a variation of an existing product. As an alternative, Bresnahan and Gordon (1997) consider that a good may be considered original if existing products are a poor substitute for it; for example horse transport is such a poor substitute for motor vehicle-based transport services that the motor vehicle can be considered a completely original product. These approaches are essentially demand-oriented. Alternatively, one may look at originality from the supply side rather than the demand side. Railways and canals both offered transport services; but, because the means by which rail travel does it are so different from the means by which canals do it, railways may be considered to have been an original product when they were first launched. Similarly, the shift from iron and steel to plastics in many industries may be thought of as a mark of originality in product innovations. Parallel definitions using a supply-based approach may also associate originality with radical innovations (loosely defined to encompass the dramatic change that transforms existing markets or industries or creates new ones) and with changes in technological trajectories (Dosi, 1982). It seems fair to consider, however, that very few product innovations will meet the necessary requirements to be classed as original under any of these definitions.

The main definition of product innovation used in data collection (especially in the CIS surveys)[3] is provided by the OECD. In the third edition of the Oslo manual, OECD (2005), product innovation is defined thus (p. 48):

> the introduction of a good or service that is new or significantly improved with respect to its characteristics or intended uses. This includes significant improvements in technical specifications, components and materials, incorporated software, user friendliness or other functional characteristics.

The definition raises several issues. First, it emphasizes significant improvements in the product as the key component of product innovation. The implication is that the greater the improvement the more significant the innovation. An alternative approach is to argue that it is not the characteristics that matter but the market impact that matters it so that a product innovation that has a greater market impact represents a more significant innovation. If one takes this to extremes, one could thus argue that a new product that does not succeed in the market, although representing a technological advance, may not be an innovation of any significance at all. The example that comes to mind is the hovercraft: it represented a revolutionary technological step, but the market for such technologies was (and is) almost zero. Extending further, there is some evidence that nearly all, or at least very many, new products introduced to most markets fail in terms of market share gain or profitability (see, for example, Castellion and Markham, 2013). Most product innovations, even if offering considerable functional improvement, may thus not be significant in market terms.

The second issue is that the OECD definition talks of innovation as involving improvements in functional characteristics. This raises three essential points. The first is that the attractiveness of a product may not be limited to functionality. In Stoneman (2011) a case is made for greater emphasis to be placed upon the role of product aesthetics as relevant market-influencing characteristics. This covers both the creative industries, where the products do not necessarily have (only) functional characteristics—for example fashion, perfumery, publishing, film, theatre—and certain manufacturing industries where aesthetic appeal may also be important—for example taste and appearance in food products, or automobile design and appearance. If such arguments are accepted, then changes in non-functional attributes may represent a significant change in the character of a product; alternatively, they may have an extensive impact upon the market success of that product and thereby represent a significant innovation.

[3] The Community Innovation Surveys is a series of surveys on innovation in different (especially European Union) countries; the series uses a largely common questionnaire and the surveys are undertaken under the auspices of the European Commission. They are a very valuable source of data on innovation activities.

The second point raised by the emphasis upon improvements in functionality is that it seems to take a view that innovation must represent product improvement in a vertical sense rather than in a horizontal sense. However, a product innovation may represent a change with respect to existing products that some buyers consider an improvement and others consider the opposite; then it can be labelled an horizontal innovation. Horizontal innovations may be of some significance. For example, (i) banking services are changing considerably, as electronic banking replaces across-the-counter banking, and these are changes of an horizontal nature that will suit some but not others; (ii) the greening of consumer products, for example changes to window-cleaning products that reduce the ability to clean windows, may again be welcomed by some buyers but not by others and are horizontal in nature (Stoneman et al., 1996). Many horizontal innovations may involve additions to product performance in one dimension but reductions in another. Such innovations are not allowed for in the OECD definition.

The third point raised by the emphasis upon functionality is that only new products that improve functionality are considered innovations. In fact this excludes a number of new products that incorporate lesser performance for lower prices, for example lower specification laptop computers for lower prices. These products may well be new and attract market success, but they are not incorporated in the OECD definition.

The implication of this discussion is that it is often assumed that product innovation means (functional) product improvement. This, however, is not necessarily true, for not only might the new product offer reduced performance but be cheaper; it may also be that the new product is horizontally differentiated (e.g. offers a non functional change) from other products. One cannot therefore judge improvement simply by exploring the extent of changes in functionality, and an alternative based upon market impact is suggested.

2.3 Industries and product innovations

As product innovation occurs, the industries in which the firms that undertake the innovation are located will develop and change over time. One might consider that some original product innovations may be so radical and game changing as to have the potential to provide the basis for the establishment of new forms of economic activity that may be classified as new industries (sometimes these are labelled 'general purpose technologies': see Bresnahan and Trajtenberg, 1995). Thus, for example, the advent of computers has led to the establishment of computer manufacturing, IT, and software sectors that were not part of the economy before the Second World War. Obviously not all the innovations, even original, will have such dramatic impact upon industrial structures; but it is upon such advances that changes in industrial structure will occur.

The new industries built upon such fundamental product innovations will have their own dynamic. The number of suppliers in the industry will change over time, the nature of the product will do the same through further product innovations, and the fundamental process technology may also change over time. In addition, as the new industry grows, old industries may well decline and eventually the new industry itself may be superseded.

The extent to which such change is reflected in the statistics will be closely related to how widely or narrowly the industries are defined. For example, the vinyl-based music-recording industry grew and then declined, being largely replaced by the CD-based music-recording industry. The CD-based music-recording industry in turn grew and then declined as downloading became the norm. However, the music-recording industry itself has continued to thrive although the products that it produces and sells have changed.

2.4 **Conclusions**

In this chapter we have introduced some basic concepts and definitions, starting with the definition of a product and distinguishing between goods and services; we have discussed product characteristics, distinguishing soft from functional; we have addressed the question of different markets for different products; and we have talked of product and market classifications. Product innovations have been described as global, local, or new to firm; new and original products have been defined; horizontal innovations have been differentiated from vertical innovations; and the issue of the significance of an innovation has been addressed. Building upon these foundations, in chapter 3 we talk of the sources from which new products are generated.

3 Sources of New Products

3.1 Introduction

The origins of product innovations are many, although some may be more important than others. Different innovations may arise in different ways from different sources. We define five main routes by which product innovations may arise in an economy: research and development (R&D), design, creative activities, emulation, and importation or foreign direct investment (FDI—or, synonymously, overseas direct investment, ODI). These are not necessarily exclusive; for example, creative activities may well be involved in design and R&D as well as being stand-alone activities. Similarly, emulation may still involve some R&D or design, and importation or FDI may also require expenditure on design or R&D to achieve adaptation. There is widespread evidence that there are strong complementarities across these different sources; for example, firms active in one direction tend to be active in others (Battisti and Stoneman, 2013; Cozzarin, 2016).

Such activities will themselves feed upon the knowledge base of the innovators. The knowledge base of a firm will be the result of that firm's own accumulated experience and abilities to access knowledge (both scientific and other), embodied in the global science base. However, the view that science always precedes technology, providing the knowledge base upon which technology feeds and creates new producers and processes, is now somewhat discredited as being too associated with a linear view of innovative activity. Instead there is currently a much more eclectic view of how science and technology interact. Technological advances are now often seen to add to the sum of total scientific knowledge, just as much as science provides a base for technological advances. Recent definitions differentiate between science and technology purely on the basis of the incentives involved. Dasgupta and David (1994) argue that scientists' and technologists' activities are governed by different incentives: the former by peer review and esteem, the latter by private profit seeking. Thus, whereas scientists undertake research in order to gain esteem through publications and dissemination, technologists seek to gain profit through secrecy and market exploitation. The implication is that science has different objectives from technology and is less market-related (see also Aghion et. al., 2008). Despite these arguments, R&D data are often still separated into basic research, applied research, and development spending, the first being associated with the science end of the spectrum, such spending being defined as having no particular application in view, the second activity being defined as research with a particular application in view, and the third being considered as expenditure incurred in moving from the prototype stage to a marketable product. In the sections that follow we begin by looking at R&D as a source of product innovations before moving to consider design, creativity, emulation, and importation or ODI.

3.2 **Research and development**

R&D spending has long been considered the prime indicator of innovative activity and as such has attracted considerable academic and statistical interest. International standards for the measurement of research and development were first put forward forty years ago in the Frascati manual. The latest edition of that manual, the third, provides a formal definition of R&D as follows (OECD, 2002, p. 30):

> Research and experimental development (R&D) comprise creative work undertaken on a systematic basis in order to increase the stock of knowledge, including knowledge of man, culture and society, and the use of this stock of knowledge to devise new applications.

Note that this definition covers basic research, applied research, and experimental development, both in science and in technology, and, through the phrase 'knowledge of man, culture and society', it includes the humanities and social sciences. R&D may be undertaken by firms, universities, government departments, international institutions, and other bodies. R&D may be financed by firms, government, or international institutions; thus R&D performed by an institution may differ considerable from R&D funded by that institution.

Clearly R&D activity is going to be a major source of new products. For example, in the pharmaceutical industries, building upon advances in chemistry, biological sciences, and biochemistry, via expenditure on R&D, new molecular entities are continually being identified as potential cures for different diseases and conditions. This enables the eventual marketing of new pharmaceutical products. In the agrichemicals business, expenditure on R&D is generating new varieties of genetically modified plants. New electronic products are being generated through expenditure on R&D, as are new telecommunications products and, for example, new motor-car engines using less fuel. It may be noted, however, that a company may not undertake all stages in the development of a new product. In pharmaceuticals, for example, it is not uncommon for small companies to proceed with a new product and, once the technological and market potential have been made clear, the company is bought by a larger operator that undertakes R&D to complete the development. In fact many years ago Jewkes et. al. (1969) presented considerable historical evidence to suggest that innovations often originated outside large firms but were then developed by large firms into marketed innovations.

However, not all R&D is directed at new products. For example, although proportions are not that clear, much will be directed at new processes rather than at products; and also some of the R&D will be basic (the proportion depending on whether one is looking at individual firms or the whole economy) and not directed at any particular target. Just as importantly, however, a number of activities that are involved in getting a product innovation to market may not be defined as R&D. It is sometimes stated that, in order to get a product to market for every £1 spent on basic research, £10 on applied research and then £100 on development are required. However, OECD (2002, p. 33) states that the definition of R&D excludes

> all those scientific, technical, commercial and financial steps, other than R&D, necessary for the implementation of new or improved products or services and the commercial use of new or improved processes. These include acquisition of technology (embodied and disembodied), tooling up and industrial engineering, industrial design n.e.c., other capital acquisition, production start-up and marketing for new and improved products.

Thus, although in 2015 the OECD economies expended 2.4 per cent of GDP on R&D (according to the OECD database: see subdirectories Science, Technology and R&D Statistics, Main Science and Technology Indicators, June, 2016) not only does not all this R&D relate to product innovation, but not all expenditure on product innovation is measured as R&D. Thus a count of R&D spending is not necessarily a good measure of product innovation activity.

To reinforce the last point, we note that OECD (2002, p. 34) states:

> The basic criterion for distinguishing R&D from related activities is the presence in R&D of an appreciable element of novelty and the resolution of scientific and/or technological uncertainty, i.e. when the solution to a problem is not readily apparent to someone familiar with the basic stock of common knowledge and techniques for the area concerned.

Thus expenditure on activities that are not devoted to either science (including the arts and the social sciences) or technological product and process innovation (excluding marketing innovations and organizational innovations) are not to be counted as R&D expenditures. To the extent that product innovation emanates from such activities, they are not adequately accounted for in the R&D statistics. The issues to which this give rise can be seen from two further quotations from the OECD manual, the first relating to the mechanical engineering industry, the second to innovation in services:

In small and medium-size enterprises...there is usually no special R&D department, and R&D problems are mostly dealt with under the general heading 'design and drawing'. If calculations, designs, working drawings and operating instructions are made for the setting up and operating of pilot plants and prototypes, they should be included in R&D. If they are carried out for the preparation, execution and maintenance of production standardisation (e.g. jigs, machine tools) or to promote the sale of products (e.g. offers, leaflets, catalogues of spare parts), they should be excluded from R&D. (OECD, 2002, p. 35)

Defining the boundaries of R&D in service activities is difficult, for two main reasons: first, it is difficult to identify projects involving R&D; and, second, the line between R&D and other innovative activities which are not R&D is a tenuous one. Among the many innovative projects in services, those that constitute R&D result in new knowledge or use of knowledge to devise new applications, in keeping with the definition [above]...In many cases, R&D findings in service industries are embodied in software which is not necessarily innovative from the technical point of view but innovates by virtue of the functions that it performs. (OECD, 2002, p. 48)

It is not often realized that the development of new software is not considered to be R&D expenditure. The outcome of this discussion is that, although R&D is a useful indicator of innovative activity, it is a flawed indicator of activity related to new product development. It includes some activities (such as new process development) that are not relevant and excludes such activities as design, which should be included if one is to capture all the relevant activities. We thus proceed to look at other activities by which product innovations may be sourced.

3.3 Design activities

Although it is not easy to find a simple or short definition of what is design (see Morris, 2009), here we consider it to involve activities that combine scientific and technical knowledge, creative flair, and possibly artistic abilities in order to produce new products and processes. In some industries, for example architecture, the output is actually a design. The second Oslo manual states that there are two important types of design (OECD, 2005, p. 41): industrial (sometimes called engineering) design and artistic design. One may think of design as midway on a spectrum between R&D and purely creative activities, in that it is not necessarily scientific or technologically path-breaking and will be more concerned with the sensory aspects of products than with their technological performance.

The Department for Business Energy and industrial Strategy (DBEIS, 2016) reports that in the 2015 UK Innovation Survey (statistical annex, Table 2) 9.9 per cent of firms reported some form of design activity, whereas 15.5 per cent of the sample firms reported internal R&D activity. This is lower than the

19 per cent reported in the 2005 UK Innovation Survey, CIS4 (see Robson and Ortmans, 2006). Using these 2005 CIS data, Tether (2006) also shows that design investment propensity does not vary significantly across industrial sectors: a similar percentage of respondents in sectors such as knowledge-intensive services and retail report design activity as in manufacturing industries. In addition, Tether (2006) finds that 71 per cent of those firms with specific design activity also have intra-mural R&D and that R&D and design tend to be deployed jointly. He finds, however, that the probability of a firm's undertaking a design function conditional on other inputs is low—that is, commitment to design seems to follow on other innovation investments, but the inverse is not the case. Although the data are far from definitive, they suggest that design activity occurs that is over and above the activities counted as R&D. Such activity occurs in every industry, although at least some of it may be associated with R&D, or even prompted by R&D. The correlation between design activity and R&D is further explored in Filippetti (2011) and in Roper et. al. (2015). It seems safe to conclude that design-based innovation plays a significant and distinctive role in innovation. We must note, however, that not all design activity is directed towards product innovation. Design, for example, may also be employed in the pursuit of process innovation. The outcome of this discussion is that, as with R&D, although a measure of the extent of design activity may be a useful indicator of innovation, it is a flawed indicator of activity related to new product development. It includes some activities (such as new process development) that are not relevant and excludes others, such as creative activity, which should be included if one is to capture all relevant activity.

3.4 Creativity

As in the case of design, finding a simple definition of creativity is not simple. In a summary of scientific research into creativity, Mumford, 2003 (p. 110) suggests that creativity involves 'the production of novel, useful products'. This is too wide for current purposes in that we are seeking to identify an activity that contributes to the creation of new products but is not encompassed per se within the R&D process or within the design process (although we are very willing to accept that creativity may form part of both).

Examples of activities that may be included are particularly relevant to the creative industries. The creative industries (see Department of Culture Media and Sport [DCMS], 2006) are considered to encompass at least the arts, media, and sports sectors—that is, the visual arts, the audiovisual arts (which include film, TV, radio, new media, and music), books and the press, heritage (which includes museums, libraries, archives, and the historical environment), sport, and tourism. However, creativity may also be of key importance in a number of other sectors such as software (not necessarily included elsewhere), architecture, finance and insurance, dining out, and so on. In industries where the main characteristic of a product is the set of its sensory characteristics (look, touch, smell, sound), creativity may well be of considerable importance.

Product innovations that are the result of creativity may include new musical performances; new musical recordings; new films; new computer games; new books; new perfumes; new gastronomic concoctions; new computer programmes; the sound of a car's exhaust; a new financial instrument; a new work of art; or a new app. The means by which such innovations are produced are heterogeneous. In the creative industries we may have the model of the lone artist in a garret who then displays works for sale in a gallery. Similarly there may be authors working at their own expense on producing new novels or other books hoping for publication eventually. Alternatively, there may be small software houses producing new computer games, or large record companies generating new music recordings, employing professional singers and orchestras. Theatre companies will invest in commissioning and then developing, rehearsing, and presenting new offerings. There may be large companies producing new films with investments in the hundreds of millions of dollars matching the investments made by drug companies in new chemical entities. Architectural practices may undertake small or large commissions meeting the expressed requirements of clients. In other industries, similarly, car manufacturers may well spend large amounts upon creative activities related to appearance, smell, colour choice, the sound made by the closing door, the shape of the grill, the angle of the headlights, and the extent of the chrome. Similarly, confectioners will devise new chocolate bars where taste will be the prime issue. In the City, investment firms and brokers may encourage financial analysts to create new financial instruments. All such creative activities may yield product innovations, but not necessarily as the result of R&D or design.

3.5 **Knowledge flows**

Whether the source of innovations be R&D, design, or creativity, a basic input into the innovation generation process is knowledge or ideas. Some such knowledge may be inherited from the past, some may be internally generated, perhaps via basic research or serendipity, and some may come from learning from current products or other activities of the innovator—since it would be a mistake to consider knowledge flows as a linear process. Of course knowledge flows may well be international. However, Peri (2005), using data on 1.5 million patents and 4.5 million citations to estimate knowledge flows at the frontier of technology across 147 sub national regions during 1975–96, estimates that only 20 per cent of average knowledge is learned outside the average region of origin, and only 9 per cent is learned outside the country of origin (although knowledge in the computer sector flows substantially farther, as does knowledge generated by technological leaders). Again, using patents as a measure of knowledge and employing a non-parametric approach based on transition probability matrices, Lamperti et. al. (2016) find positive impacts of knowledge flows from developed countries (OECD) on innovative activities in Africa, at the firm level.

Interestingly, on the basis of an econometric analysis of a matched data set comprising innovation panel data and plants' patent histories for Irish firms, Roper and Hewitt-Dundas (2015) find evidence that existing knowledge stocks—measured by patents—have only a weak negative impact rather than a positive one on firms' innovation outputs. Perhaps this reflects potential core rigidities or negative path dependencies. Instead knowledge flows derived from R&D investment and external search dominate the effect of existing knowledge stocks on innovation performance. The key may therefore be to seek the sources of contemporaneous knowledge.

Over the last decade there has been particular emphasis in the literature on the concept of open innovation (see, for example, Chesbrough, 2003). This concept or paradigm emphasizes that that firms can and do use external as well as internal ideas as a means to advance their technology. One would thus expect to find both knowledge inflows and outflows across organizational boundaries, influencing the company's knowledge stock. One implication of the open innovation paradigm is that inflows and outflows of knowledge are complementary. Cassiman and Valentini (2016), however, find no evidence for such complementarity in a sample of Belgian manufacturing firms. Firms buying and selling knowledge do increase their sales of new products, but at the same time their R&D costs increase more than proportionately.

Clearly there are many sources of information inside and outside the innovating organization. The UK Innovation Survey, 2015 (for which see the Department for Business, Innovation and Skills [DBIS], 2016) asked respondents, over a period from 1 January 2012 to 31 December 2014, how important the information from 12 different sources was to the business' innovation activities, requesting a ranking—high, medium, low, or not applicable—for each source. The sources encompass (i) your business or enterprise group; (ii) suppliers of equipment, materials, services, or software; (iii) clients or customers from the private sector; (iv) clients or customers from the public sector; (v) competitors or other businesses in your industry; (vi) consultants, commercial labs, or private R&D institutes; (vii) universities or other higher education institutes; (vii) government or public research institutes; (ix) conferences, trade fairs, exhibitions; (x) professional and industry associations; (xi) technical, industry, or service standards; (xii) scientific journals and trade/technical publications standards.

Table 3.1 Sources of information for innovation

	Within your business or enterprise group	Suppliers of equipment, materials, services, or software	Clients or customers from private sector	Clients or customers from public sector	Competitors or other businesses in your industry
All	46.7	23.3	19.9	8.8	13.3
Size of enterprise					
10–49	45.4	22.8	18.8	8.5	13.2
50–99	50.0	27.0	24.5	9.2	12.8
100–249	52.0	22.8	23.4	10.8	14.3
250+	59.7	24.7	26.0	12.1	15.6
High/low-technology industries					
High-tech manufacturing	56.0	18.8	29.0	11.1	13.7
Low-tech manufacturing	51.7	26.0	22.8	8.2	12.8
Other industries	45.0	23.3	18.6	8.7	13.4

Source: DBIS (2016), *UK Innovation Survey 2015*, Statistical annex, Table 9. (https://www.gov.uk/government/statistics/uk-innovation-survey-2015-statistical-annex-and-interactive-report)

In Table 9 of the statistical annex to the UK Innovation Survey 2015, the sources of information (measured by the percentage of all the firms with some innovation activity that rank the source as being of 'high' importance) are listed, from which we tabulate the data relating to the five most important sources in Table 3.1.

The results relate to both product and process innovation, but one does not expect systematic differences between these two activities. These results confirm that both internal and external sources are important, internal sources being of high importance to about half of the innovating firms. The proportion of innovating firms that consider internal sources important increases for larger firms and high-tech industries. Of the external sources, the main sources considered important by innovating firms are suppliers of equipment and clients or customers from the private or public sector: these two, jointly, form the most highly rated external source. Finally, competitors are also ranked as an important source, but not excessively so. One may also note that fewer of the smallest innovating firms rank these sources as being of high importance. Surprisingly, universities (1.8 per cent), government and public research institutes (1.9 per cent), and scientific publications (1.5 per cent) do not figure prominently as being of high importance to innovators. Consultants (4.4 per cent), conferences (5.5 per cent), and professional and industry associations (5.6 per cent) figure more strongly, as do technical, industry, or service standards (5.8 per cent).

Of course, the sourcing of information is not a process that occurs like manna from heaven. Organizations will incur expenses in sourcing knowledge from outside, in integrating that knowledge with their internally sourced knowledge, and then in effectively exploiting it. There is literature in economics on the undertaking of search activity, recent literature encompassing equilibrium models of search (see, for example, McCall, 2014). Similarly, there is a literature in management on knowledge management per se (for example, du Plessis, 2007 and Tidd, 2012).

3.6 Imports and FDI

An alternative source of product innovations for a national economy is imports. Instead of being produced at home, they may be sold on the domestic market by an overseas supplier. In fact, although product innovations may be generated, produced, and sold in the same country or region, they are more generally invented in one country, developed in another, and launched and produced in one or several different countries.

In some markets where there are no local production facilities product innovations will have to be imported. Thus for example in the United Kingdom no televisions, computers, mobile phones, or laptops are produced. Any product innovations in these markets must be imported. However, even where there are local production facilities, those domestic producers may not be the lead innovators and product innovation may still occur via importation. Of course, product innovations produced in the domestic economy may be exported.

Over time it may be that overseas producers establish production facilities in the domestic economy via FDI (or ODI), which may mean that new products are produced in the domestic economy (and in such cases the date of first production of the new product in the economy will be later than the date of product innovation in that economy by that supplier).

To some degree, therefore, product innovation in an economy will depend upon the exporting decisions of overseas suppliers. The FDI decisions of such suppliers may also determine the extent to which product innovations are supplied from the domestic economy.

3.7 Emulation

Although some R&D, design, or creative effort may still be involved, the emulation or copying of others is a further means by which new products can be sourced. Such copying may dilute the returns to the originator but stimulates competition in the market and hence benefits the consumer. Of course emulation will only produce product innovations that are new to the firm and not products that are new to the market.

It is not easy to measure the extent to which product innovation reflects emulation rather than self-origination. Dacko et al. (2015) undertake an empirical analysis using data on the timing behaviour of ninety-nine 'non-pioneering' or following firms that introduced low-fat products into US food markets. Preliminary analysis indicated that in aggregate, the mean duration between the pioneer's launch and the followers' launch of a new product was 45.8 months, with a standard deviation of 23.4 months. Following seems common, but the timing of followers reflects wide disparity. However, not all the followers should be considered emulators. The data illustrate that the product development actions of many follower firms temporally preceded the actual pioneering product launch. Of the ninety-nine follower launches observed, in twenty-four cases the firms' attention had already been attracted by the potential of the new product at the date of the pioneer launch in their industry; fifteen of these firms had initiated market research, and thirteen had already initiated new product development; only seventy-five did not consider the new product until after pioneering product launch. Apparent emulation may thus in fact be the result of follower firms acting prior to the actual pioneering launch.

Emulating may be less risky than being a path-breaker, for the leader may well have proved that the market exists and that the product is technologically viable. However, there may be first-mover advantages that make being second (or later) less attractive than it might otherwise be (for an early discussion, see Van der Werf and Mahon, 1997). Lieberman and Montgomery (1988) summarize the reasons for the existence of early-mover advantages: proprietary learning effects, patents, pre-emption of input factors and locations, and development of buyer switching costs. Conversely, they also argue that early-mover disadvantages may result from free-rider problems, delayed resolution of uncertainty, shifts in technology or customer needs, and various types of organizational inertia. The essence of early-mover advantages is that an early mover may have the potential to affect the decisions of later movers. In an early study, Lambkin (1988) shows that pioneers tend on average to outperform later entrants, but by comparison to early followers and late entrants have significantly different strategic profiles and performance levels. Schnaars (1994) on the other hand argues that being an imitator is the best route. In a much more recent study, Lanzolla et. al. (2010) show that in the European mobile telephony market over the period 1998 to 2007 first and second movers consistently outperformed later entrants.

However, given that the launch of competing new products on a market may restrict the returns to the originator, emulation may be limited by both formal and informal intellectual property protection. For example, in addition to the use of the formal mechanisms that we discuss, Geroski (1995) argues that originators may also try to limit emulation informally through secrecy, information-sharing arrangements, control of strategic resources, specialized knowledge, threats, and learning by doing. It should be noted, however, that limiting emulation by such formal and informal methods may have other wider impacts; for example, by protecting the original supplier of a new product, it can deter research on and the development of different but similar product innovations. It should also be noted that (as it is argued), by disseminating the technological information contained in a formal grant of intellectual property, one may assist future technological development.

Intellectual property rights (IPR) are designed to protect the rights of the originator, inventor, or pioneer of a new technology by discouraging imitation. The four main institutional arrangements that exist in most countries in order to reinforce IPR are patents, which protect the technical and functional aspects of products and processes; design rights (or design patents), which protect the visual appearance or eye appeal of products; copyright, which protects material such as literature, art, music, sound recordings, films, and broadcasts; and trademarks, which protect signs that can distinguish the goods and services of one trader from those of another. Each has its own particular modus operandi.

Patents protect new inventions and cover how things work, what they do, how they do it, what they are made of, and how they are made. They give the owner the right to prevent others from making, using, importing, or selling an invention without permission. In many countries only advances of an industrial nature can be patented, which causes them to be labelled utility patents. In other countries, such as the United States, there is also a form of patents called design patents. These correspond to the design rights that offer protection elsewhere. To be granted a utility patent, an invention must be new, must involve an inventive step that is not obvious to someone with knowledge and experience in the subject, must be capable of being made or used in some kind of industry, and must not be a scientific or mathematical discovery, theory, or method, a literary, dramatic, musical, or artistic work, a way of performing a mental act, playing a game, or doing business, the presentation of information, or a certain type of computer programs, an animal or plant variety, a method of medical treatment or diagnosis, or against public policy or morality. There will of course be many product innovations that fall outside these restrictions and therefore are not patentable.

Once granted a patent, the holder has the right to (not unfairly) determine access to the knowledge embodied in the patent and charge licence fees for use of that knowledge. If granted, the patent must be renewed every year after the fifth and may then provide protection for up to twenty years. If the holder becomes aware that his/her rights are being infringed, there is recourse to the courts, with the possibility of terminating the infringements and receipt of damages.

Within the EU, firms may apply for either national or European patents, the latter proving community-wide protection, the former only national protection. Of the applicants at the European Patent Office (EPO; not all for product as opposed to process innovations), the top twenty-five applicants represent about one fifth of the patent applications filed and 1 per cent of the applicants receive more than fifty patent grants per year (and thus applications are concentrated in a few firms), but there is a long tail in that above two thirds of the applicants at the EPO are granted only one patent per year.

Design rights apply to intellectual property but relate to the physical appearance of a product. They are not concerned, like patents, with the function or operation of that product, emphasizing instead the appearance that results from the features of physical products or the way they look. Design rights thus may offer protection to product innovations that would not be covered by patents; or they may provide additional protection to the design aspects of a new product over and above that offered by patents.

In different countries there are different types of design protection available. For example, registered designs offer protection throughout the United Kingdom. Application must be made for this IPR, a fee has to be paid, and it is not an automatic right. registered community design (RCD) offers similar protection in all of the EU member states. UK design right is an automatic right that does not need to be applied for and prevents others from copying a design; but it covers only the 3D aspects of the item and does not protect the surface decoration of the product or any 2D pattern such as a wallpaper or a carpet design. The unregistered community design right is also an automatic right for which one does not need to apply and offers protection from copying the design on any item.

To qualify for any of these rights, the design must be new and individual in character, which means that the overall impression the design gives the informed user must be different from that of any previous designs. A design cannot be registered if it is more than twelve months since it was first publicly disclosed; if it is dictated only by how the product works; if it includes parts of complicated products that cannot be seen in normal use (for example, vehicle engine spare parts, or the parts inside a computer); if it is offensive; or if it involves certain national emblems and protected flags. A registered design provides the right to sell or licences someone else to use it.[1]

In Table 3.2 we present some data on the top ten companies applying for design registrations in the EU in 2011. This is similar to patterns found in other countries. The World Intellectual Property Organisation (WIPO, 2012) concludes that the top applicants show that the electronics and ICT, automotive, clothing and fashion, interior design, and decoration industries—and, to

[1] Examples of protected designs may be found by looking at www.ipo.gov.uk/design/d-applying/d-should/d-should-designright.

Table 3.2 Leading applicants for design registration, OHIM, 2011

Rank	Company	No. of applications	Sector
1	Rieker Schuh AG	947	Footwear
2	Microsoft Corporation	644	IT
3	Electrolux Home Products Corporation	500	Electricals
4	Sony Corporation	485	Electricals
5	Eglo Leuchten Gmbh	476	Lighting
6	Pierre Balmain, Societe Anonyme	437	Fashion
7	Creation Nelson	403	Fashion
8	Samsung Electronics	350	Electrical
9	Nike International Limited	319	Sportswear
10	Koninklijke Philips Electronics	318	Electrical

Source: WIPO (2012).

a lesser extent, firms in the consumer product industries—use the industrial design rights system most intensively.

Copyright is a third IPR. Copyright relates to the expression of an idea, not to the idea itself or to any process by which that idea is embodied in a physical artefact. The UK Intellectual Property Office (IPO) illustrates this right by citing the example that anyone can write a story based on the idea of a boy wizard, but they cannot copy text or illustrations (without permission) from other books on the same subject. This is very much in contrast to the patent system, where the idea itself is protected and owned for a period by the patent holder. But it does not protect the names, designs, or functions of the items themselves.

Copyright is particularly applicable to new products in the creative industries. Copyright protects sound recordings, films, broadcasts, and original artistic, musical, dramatic, and literary works, including for example photographs, sculptures, websites, computer programs, plays, books, videos, databases, maps, and logos. But it does not protect the names, designs, or functions of the items themselves. For example, copyright is the mechanism that may be used to protect originators from the unauthorized downloading of music and films from the internet or against copying from other originals. Software enjoys only copyright protection and not patent protection.

It is not necessary to formally apply or pay for copyright in the United Kingdom.[2] It is an automatic right. The copyright arises as soon as the work is 'fixed', for example written down, recorded, or stored (in a computer memory); and in the United Kingdom it is established once the © symbol is attached to the work with the creators' name and the date when that work was created. However, as copyright is not registered in the United Kingdom, it

[2] Although this is not the case in all countries.

is particularly difficult to obtain numbers of the copyrights claimed. The owner of the copyright has the right to license, sell, or otherwise transfer that copyright to someone else.

Copyright in literary, musical, artistic, and dramatic work in the United Kingdom lasts for the creator's lifetime plus seventy years (basically the same as in the EU and in the United States). For films it is seventy years after the death of the last of the directors, score composer, dialogue or screenplay authors; and for TV and radio programmes it is fifty years from the first broadcast. Sound recording copyright lasts for fifty years. Publisher's right, which covers the typographical layout of published editions such as books or newspapers (how they are presented on the page), lasts for twenty-five years from creation. These lives are considerably longer than the terms of even extended patent rights.

The fourth basic IPR mechanism is the trademark. A trademark is a sign that distinguishes a firm's goods and services from those of other traders. A sign includes, for example, words, logos, pictures, or a combination of these. Whereas patents require novelty and copyright requires originality, the counterpart for trademark is distinctiveness. Whereas patents are not available for aesthetic innovations, such innovations may be trademarked. Non-aesthetic innovations may also be trademarked. For example, a rock group can trademark its name, a product with a particular aesthetics, such as the iPAD, can be trademarked, and particular products such as Mars bars and Crunchie bars may also be trademarked.

A registered mark confers (1) the right of use of that mark on the goods and services in the classes for which it is registered and (2) the legal right to take action against anyone who uses that mark or a similar one on the same or similar goods and services—similar, that is, to those that are set out in the registration. To be registrable, the trademark must be distinctive for the goods and services for which application is made and must not be the same as, or similar to, any earlier marks on the register for the same or similar goods or services.

A trademark does not have to be registered. An unregistered trademark provides certain rights under common law and the owner can use the TM symbol. However, it is easier to enforce rights if the mark is registered, which also enables use of the ® symbol to indicate registration. In the United Kingdom application for registration is made to the Trade Marks Register of the IPO. There is a need to pay a renewal fee every ten years. European protection via a Community Trade Mark application is made through the European Union Intellectual Property Office (previously the Office for Harmonization in the Internal Market). As with other IPR instruments, it is necessary for the owner to police his/her rights in person, via the courts.

In Table 3.3 we present some data indicating the use of patents and trademarks worldwide within the context of world R&D spending. Note, however,

Table 3.3 R&D spending and IP applications worldwide, 2002–11

Year	R&D Spending (GERD) $Bn. ppp	Patent applications	Trademark applications	Industrial design applications
2002	787.7	1442500	2389889	322039
2007	1155.4	1866700	3328561	520837
2009	1276.9	1846700	3253887	586785
2010	–	1985300	3686502	668470
2011	–	2140600	4175987	775631
% growth 2002–2009	62	28	36	82
% growth 2010–2011	–	7.8	13.3	16.0

Sources: R&D from UNESCO Institute for Statistics [UIS] (2012a), remainder from WIPO Statistics Database, October 2012.

that these figures also encompass aspects of innovation other than product innovation. We do not have comparable data on copyrights issued. Although counts of design right applications may reflect changing behaviour with respect to registration rather than increased levels of design activity (a problem we are unable to overcome), the growth rates presented in Table 3.3 indicate that design right applications have grown faster than the use of the other intellectual property mechanisms. There are also indications that trademark applications grew faster than patent applications. One may observe that design right applications have been growing faster than the total R&D spend.

Overall, therefore, whereas a number of product innovations may be patented, many others are not. However, these others may be capable of being copyrighted or trademarked, or they may even employ a design right protection as well. It may also be the case that some innovations are protected by one or more of these mechanisms as well as by less formal barriers that will slow or prevent emulation. These factors may well slow the extent of emulation. It may also be the case that these factors impact upon other sources of product innovations, such as R&D and design activity, as well as upon pure copying or emulation.

3.8 **Conclusions**

In this chapter we have discussed the sources of product innovations and presented five main sources: R&D, design activity, creativity, imports (and foreign direct investment), and emulation of others. We are not able to provide a neat statistical breakdown of the relative importance of the various sources, but it is worthy of note that, despite the views of many commentators,

R&D is not the sole source of new products. It may be especially important in some industries, but in others—such as the creative industries—design and creativity may be more relevant. Also, for any individual economy, many product innovations may originate or be supplied from outside the national economy. One might also think that emulation is a major source of innovative new products, although we have discussed how emulation can be limited by formal and informal means of protecting intellectual property.

4 The Extent of Product Innovation

4.1 Introduction

In this chapter we present and analyse data on the extent of revealed product innovation in different economies and sectors. The most commonly used indicators of innovative activity are R&D and patenting or other intellectual property rights (IPR) activity, but we consider that (i) the former is an (incomplete) measure of inputs to the innovation process, covering both products and processes, and not a good measure of the extent of product innovation; and (ii) the latter measures encompass many products (and processes) that never make it to market, while many new products are not formally protected, which makes such indicators unreliable (see Griliches, 1990 and Nagaoka et. al., 2010 for discussion of the use of patent-based measures as indicators of the extent of innovation). For these and other reasons discussed in chapter 3, most of the data we analyse here relate directly to the number of new products launched in different economies and sectors. We also attempt to identify the extent to which product innovations succeed in establishing themselves upon the market. Basically, those that have little market impact may be of little economic importance—although this not to deny that some cultural products that are not widely diffused may have impact in other ways, a matter discussed elsewhere (Stoneman, 2011). We start by looking at data from innovation surveys in various economies before exploring other product launch data. As such data tend to provide snapshots of the world, which could be misleading, we finally explore whether industry product innovation patterns reveal a life cycle over which the tendency of firms to undertake product innovation varies with the maturity of the industry in which these firms are located.

4.2 Innovation surveys

4.2.1 *Europe*

As part of an exercise coordinated by Eurostat, the individual statistical offices in the different EU member states have, for a number of years, been collecting data about innovation activity in the different states via the community innovation surveys (CIS). The data we present here cover surveys undertaken from 2004 until 2012. A survey generally refers to innovation activity in the three years prior to its date. We concentrate upon data that reflect the propensity of sample firms to undertake product innovation and how this differs across country, sector, firm size, and time. CIS and other surveys measure not only the propensity to product innovate but also the propensity of firms to undertake other types of innovation, such as process innovation, organisational or managerial innovation, and marketing innovation. As has been shown by Battisti and Stoneman (2010), there is a high correlation between a firm's undertaking one type of innovation and its undertaking other types of innovation. Thus firms that are innovative in one dimension (for example product innovation) tend also to be innovative in other dimensions (such as process innovation and or managerial innovation), a matter considered in some detail in chapter 8. Although such complementarities may be part of the driving force behind product innovation, in this chapter we are primarily concerned with the pattern of product innovation alone, and therefore we measure primarily whether firms undertake product innovation or not rather than whether such innovating is accompanied by other kinds of innovating.

In Table 4.1 we present data relating to the latest (2012) Europe-wide surveys (CIS 8) on the percentage of sample firms in thirty-one countries that have undertaken product innovation during the period of the survey (these data are under the label 'productive innovative enterprises regardless of any other type of innovation', downloaded from the Eurostat data base on 11 November 2016; see appso.eurostat.ec.europa.eu/nui/show.do?dataset = inn cis8). Of the thirty-one countries, Norway, Serbia, and Turkey are not part of the EU, while the fifteen countries that were members in the European Union prior to the accession of the ten other candidate countries on 1 May 2004 are Belgium, Denmark, Germany, Ireland Greece, Spain, Francc, Italy, Luxemburg, the Netherlands, Austria, Portugal, Finland, Sweden and the United Kingdom.

The data indicate that in the sample period approximately a quarter of all firms undertook a product innovation. Germany, where 35.8 per cent of the firms were product innovators, Finland, Sweden, Belgium, and the Netherlands were the most active, whereas Romania (3.4 per cent), Poland, and Latvia were the least active.

Within these totals, an interesting breakdown is between innovations in goods and innovations in services. However, this is not easily accessible for the 2012 survey. Table 4.2 is thus based on the 2010 data in CIS 7. The data in the first and second columns indicate the proportion of firms that introduced new goods and the proportion that introduced new services (with or without organizational or marketing innovation). However, as firms may do both, the sum of these two series may overestimate the proportion of firms that

Table 4.1 The percentage of firms that have undertaken product innovation, 2012 CIS surveys

State	% of enterprises
European Union (28 countries)	23.7
European Union (15 countries)	26.9
Belgium	31.5
Bulgaria	10.8
Czech Republic	25.3
Denmark	24.8
Germany (until 1990 former territory of the FRG)	35.8
Estonia	20.7
Ireland	27.8
Greece	19.5
Spain	10.5
France	24.2
Croatia	16.4
Italy	29.1
Cyprus	20.9
Latvia	10.4
Lithuania	11.6
Luxembourg	30.3
Hungary	10.6
Malta	23.9
Netherlands	31.9
Austria	26.6
Poland	9.4
Portugal	26.0
Romania	3.4
Slovenia	23.6
Slovakia	14.4
Finland	31.0
Sweden	31.5
United Kingdom	24.0
Norway	19.1
Serbia	24.5
Turkey	17.7

Source: Eurostat (2016), eurostat.ec.europa.eu/nui/show.do?dataset = inn cis8.

undertake product innovation. These latter data have only limited geographic coverage but clearly show that the proportion of firms introducing new goods varies from 32.9 per cent in Iceland through to 7.4 per cent in Latvia, while the proportion of firms introducing new services varies from 29.5 per cent in Luxembourg to 3.5 per cent in Latvia. In thirteen of the twenty-eight countries

Table 4.2 The percentage of sample firms innovating in goods and services, 2010 CIS 7

State	Innovation, new goods	Innovation, new services
Belgium	26.6	18.1
Bulgaria	8.5	3.6
Czech Republic	19.4	11.3
Estonia	17.4	12.6
Ireland	20.5	14.5
Spain	11.2	5.9
France	18.8	12.5
Croatia	16.1	9.6
Italy	24.5	14.4
Cyprus	15.9	16.8
Latvia	7.4	3.5
Lithuania	8.5	7.9
Luxembourg	27.0	29.5
Hungary	9.7	5.2
Malta	12.1	11.1
Netherlands	25.2	18.0
Austria	26.1	15.3
Poland	8.0	4.1
Portugal	22.9	18.2
Romania	7.6	6.2
Slovenia	20.7	12.0
Slovakia	16.3	7.4
Finland	23.8	17.9
Sweden	25.7	8.8
Iceland	32.9	26.0
Norway	16.2	7.2
Serbia	22.7	20.9
Turkey	20.4	15.9

Source: Eurostat, Enterprises by type of innovation, CIS 7, last updated 12 March 2014 (appso.eurostat.ec.europa.eu/nui/show.do?dataset = inn cis7).

for which we have data, more than 20 per cent of the firms report introducing new goods, but in only three countries do more than 20 per cent of firms report introducing new services (although 15 report that more than 10 per cent of firms do so). No obvious pattern is revealed.

Using data from several of the CIS surveys, we may also explore whether product innovation has been increasing or decreasing over time. However, the simple indicator for the 2012 survey of the extent of product innovation (with or without other innovation activities) is not available for other years, and thus in Table 4.3 we detail the number of firms that undertake either product

Table 4.3 Proportion of sample firms undertaking product innovation (%), CIS, 2004–10

State	2010	2008	2006	2004
Germany	41.5	41.2	45.9	43.3
Spain	13.7	14.3	16.2	18.7
France	37.3	23.4	–	19.4
Italy	28.7	27.4	18.7	18.4
Netherlands	34.9	24.8	25.3	24.5

Source: Calculated from data available from Eurostat, Science Technology and Innovation database, accessed 23 June 2014, own calculations.

innovation or both product and process innovation (thus excluding those that also undertake organizational or marketing innovation) as a proportion of the total number of enterprises in the population for Germany, Spain, France, Italy, and the Netherlands in the CIS for 2004, 2006, 2008, and 2010. As the 2012 data are not strictly comparable, we do not directly make a comparison. The data show different trends in different countries. The proportion of firms undertaking product innovation has been increasing in France, Italy, and the Netherlands, but decreasing in Germany and Spain.

The data indicate a high level of product innovation among European firms. The next matter of interest is whether this product innovation is new to the firm or new to the market. One should note that individual firms may have done both. In Table 4.4 we again use data from CIS 7 (i.e. 2010) to indicate this. In column A we show the number of firms that have introduced goods (thus services are excluded) that are new to the market relative to the number that have introduced goods that are new to the firm (as a percentage). In the CIS surveys there also a number of questions relating to the proportion of the firm's sales during the survey year that derives from products that are new to the firms in the three years prior to the survey. This indicator reflects to various extents the amount of product innovation undertaken and also the market success of such innovation (which in turn may be dependent, inter alia, upon innovation by others, costs, pricing, and marketing activity). Although this may not be a pure indicator of the extent of product innovation, it is useful for current purposes to reflect the extent of innovation that is new to market relative to innovation that is new to firm (column A). In column B of Table 4.4 we show, for the product-innovating firms, the turnover arising from product innovation that is new to the market as a percentage of the turnover arising from product innovation that is new only to the firm.

These data indicate that, except for Iceland and Italy, there are fewer firms introducing products that are new to market than firms introducing products that are new only to the firm itself. There is, however, considerable dispersion across the sample states. The ratio varies from a high of 121.2 in Iceland

Table 4.4 Product-innovating firms: goods new to firm and new to market, 2010 CIS 7

State	A Innovating enterprises, ratio new to market/new to firm %	B Turnover from product innovation, ratio new to market/new to firm %
European Union (28 countries)	–	53.7
European Union (15 countries)	–	50.6
Belgium	86.3	93.0
Bulgaria	62.9	104.7
Czech Republic	77.0	94.8
Denmark	85.5	93.8
Germany (until 1990 former territory of the FRG)	48.4	30.9
Estonia	56.5	51.4
Ireland	–	89.1
Spain	60.8	87.9
France	87.9	30.7
Croatia	72.7	60.4
Italy	101.4	103.6
Cyprus	58.5	75.9
Latvia	82.1	86.7
Lithuania	75.1	58.8
Luxembourg	69.7	171.1
Hungary	86.2	205.7
Malta	85.0	98.8
Netherlands	87.9	103.8
Austria	90.3	76.1
Poland	87.6	132.0
Portugal	63.5	104.5
Romania	51.8	45.6
Slovenia	91.8	83.4
Slovakia	82.9	139.1
Finland	77.9	122.2
Sweden	96.5	101.8
United Kingdom	68.4	35.7
Iceland	121.2	69.8
Norway	79.9	112.2
Serbia	60.1	56.5
Turkey	–	122.7

Source: Eurostat, CIS 7, Product and Process Innovation, last updated 25 March 2014.

through 101.4 in Italy down to 48.4 in Germany. The unweighted mean is 90.2. Although we see that the firms introducing product innovations that are new to the market are fewer than those introducing innovations that are new to the firm, the second column illustrates that in a number of states the new-to-market innovations undertaken contribute more to turnover than the new-to-the-firm innovations. This is the case for example in Hungary (205.7 per cent), Luxembourg (171.1 per cent), Finland (132 per cent), and Poland (132.0 per cent). We notice, however, that only in Italy do both the number of new-to-market innovations introduced by enterprises and the turnover deriving from these innovations exceed 100 per cent of equivalent new-to-the-firm numbers. This suggests that the large ratios of turnover derived from innovations new to market by comparison to innovations new to firm do not necessarily arise from there being more of the former than of the latter.

One may explore CIS data further by both firm size and industrial sector. However, the task of presenting detail for all the separate EU countries is a difficult one. We thus concentrate on the UK pattern. This also has the advantage that the UK data are more recent than data for the EU as a whole (the latest survey was in 2015). The overall pattern of product innovation in the United Kingdom is shown in Table 4.5.

These data indicate a rate of product innovation (measured by the proportion of firms product-innovating) with the following characteristics: (i) the proportion of product innovating firms (about 20 per cent) exceeds the rate of firms process innovating firms (about 11 per cent); (ii) the rate of product innovation has been slowly declining from 24 per cent to 18 per cent over the period from 2009 to 2015; (iii) for large firms, the rate of product innovation is greater than for smaller firms; and (iv) the proportion of product innovations that are new to market is greater in larger firms, but this proportion has recently declined to about one in three in all firms. Additionally, the average share of total turnover in 2014 arising from product innovation new to market was 14.4 per cent and from new to the firms was 16.3 per cent (DBEIS, 2016, Table 4).

Table 4.5 Product innovation in the United Kingdom, 2009–15 innovation surveys

	2009	2011	2013	2015
% product-innovating firms (all)	23.9	18.9	18.0	18.4
of which new to market	–	–	44.0	31.0
% product-innovating firms (250+ employees)	–	–	23.0	27.0
of which new to market	–	–	50.0	39.0
% product-innovating firms (10–250 employees)	–	–	18.0	19.0
of which new to market	–	–	44.0	31.0
% process-innovating firms (all)	12.6	10.3	10.5	11.9

Sources: DBEIS (2016), Interactive Innovation Activity Dashboard (http://analysis.bis.gov.uk/ukinnovationsurvey2015/dashboard, accessed 11 November 2016); UK Innovation Survey 2015.

In Table 4.6 we present the detail of the UK product innovation breakdown by industrial sector. The UK data have the advantage that they cover a wide set of industries and, unlike some CIS surveys in other countries, include some of the creative industries. Although we have some doubts as to how well the CIS questionnaire picks up product innovation in the creative sector, the inclusion is still an advantage. The sample for CIS data of this kind is restricted to firms with more than ten employees.

The rates of product innovation refer to the period 2012–14, whereas the division between turnover arising from product innovations that are new to market and from those new to firms is for 2014. The first data presented confirm the overall pattern of product innovation presented in Table 4.5, but with greater detail across firm sizes. These data indicate that the rate of product innovation increases with firm size but the share of turnover that is due to innovations that are new to market declines with firm size. One should note that innovation by start-up firms would not be captured by the CIS methodology.

The high/low-tech split suggests that in manufacturing high-tech firms are more product-innovative than low-tech firms and have a greater proportion of turnover that is due to product innovation that is new to market. Other industries appear to be less product-innovative than either type of firm in manufacturing and a greater share of their turnover is due to product innovations that are new to business.

By broad industrial sector, more firms undertake product innovation in engineering-based manufacturing (at 43.8 per cent) and in knowledge-intensive services (at 31.5 per cent), while fewer are innovative in other services (14.8 per cent) and in construction (13.0 per cent). New-to-market innovations contribute the most to turnover in knowledge-intensive services (23.3 per cent) and least in construction (9.4 per cent), whereas new-to-business ones contribute most to turnover in construction (23.1 per cent) and least in other manufacturing (13.1 per cent). At a finer level of industrial detail we observe more considerable differences across industries in the proportion of firms that undertake product innovation: this varies from a low of 8.6 per cent in renting of machinery to a high of 45.2 per cent in the manufacture of electrical and optical equipment, 39.9 per cent in computer and related activities, and 38.3 per cent in research and experimental development. The importance of product innovations in total turnover also differs considerably by industry at this low level of aggregation. The contribution to turnover of new-to-market innovations is greatest in mining and quarrying (53.9 per cent); next come other professional, scientific, and technical activities (49.7 per cent), research and experimental development (37.2 per cent), and computer and related activities (27.1). The contribution is lowest in real estate activities, clinical testing and analysis (4.2 per cent), and financial intermediation (5.7 per cent). The contribution to turnover of new-to-business

Table 4.6 UK enterprises engaging in product innovation, 2012–14, and shares of total turnover arising from product innovation, 2014

	Product innovator % sample firms	New to market % turnover	New to business % turnover
All	19.2	14.4	16.3
Size of enterprise			
10–49	18.4	15.3	17.0
50–99	21.4	10.8	13.4
100–249	23.2	10.4	14.7
250+	26.8	10.2	12.0
High/low-technology industries			
High tech manufacturing	33.7	16.3	13.7
Low tech manufacturing	29.9	9.9	13.1
Other industries	16.8	15.2	17.4
Broad sectors			
Primary sector	14.8	18.2	16.4
Engineering-based manufacturing	43.8	14.6	13.6
Other manufacturing	28.5	10.3	13.1
Construction	13.0	9.4	23.1
Retail and distribution	16.0	11.7	12.5
Knowledge-intensive services	31.5	23.3	18.1
Other services	14.8	14.4	19.3
Sampling sectors			
Mining and quarrying	13.1	53.9	20.6
Manufacture of food, clothing, wood, paper, publishing and print	29.5	10.2	12.5
Manufacture of fuels, chemicals, plastic metals and minerals	28.2	10.3	12.6
Manufacture of electrical and optical equipments	45.2	13.3	13.0
Manufacture of transport equipment	37.5	24.9	16.5
Manufacturing not elsewhere classified	27.0	11.0	15.3
Electricity, gas and water supply	15.1	13.9	16.0
Construction	13.0	9.4	23.1
Wholesale trade (including cars and bikes)	18.5	11.1	14.3
Retail trade (excluding cars and bikes)	12.8	12.6	9.5
Transport	11.7	17.4	19.0
Post and courier activities	17.4	16.7	7.2
Hotels and restaurants	13.8	5.8	15.5
Computer and related activities/ICT	39.9	27.1	16.3
Motion picture, video and TV programme, production/programming and broadcasting	29.0	19.8	13.7
Telecommunications	30.0	20.8	19.4
Financial intermediation	19.9	5.7	22.6
Real estate activities	12.8	0.5	68.9

Other services not elsewhere classified	14.0	16.9	18.3
Architectural and engineering activities and related technical consultancy	27.8	15.2	18.3
Clinical testing and analysis	26.9	4.2	13.1
Research and experimental development on social sciences and humanities	38.3	37.2	19.0
Advertising and market research	30.8	12.7	10.0
Other professional, scientific and technical activities	33.0	49.7	14.7
Renting of machinery, equipment, personal and household goods	8.6	9.5	24.0

Sources: DBEIS (2016), *The UK Innovation Survey 2015*, Statistical annex, Tables 1 and 5 (https://www.gov.uk/government/statistics/uk-innovation-survey-2015-statistical-annex-and-interactive-report, accessed 15 November 2016).

innovations is greatest in real estate activities (68.9 per cent), construction (23.1 per cent), and financial intermediation (22.6 per cent) and lowest in post and courier activities (7.2 per cent), retail trade (9.5 per cent), and advertising and market research (10.0 per cent). It is also worth pointing out that, in two sectors that are not often considered in studies of innovation, financial intermediation and the creative industries, the data do show significant levels of product innovation activity. Thus, for example, in financial intermediation 19.9 per cent of firms undertake product innovation and, although only 5.5 per cent of turnover is due to new-to-market products, some 22.6 per cent is due to new-to-business product innovation. For an analysis of the UK innovation survey data regarding innovation in the finance sector, see Heffernan et al. (2013). The creative industries cover a number of sampling sectors. Taking hotels and restaurants, architectural services, motion picture and video production, and advertising as representative of the creative industries, we observe that between 13.8 per cent and 30.8 per cent (depending on the sector) of sample firms are product innovators, new-to-market turnover being between 5.8 per cent and 19.8 per cent and new-to-business turnover between 7.2 per cent and 18.3 per cent. Overall, the percentages of firms doing product innovation and the market shares deriving from such innovation in financial intermediation and the creative industries are not out of line with the more frequently studied manufacturing sector.

4.2.2 *The United States*

Earlier versions of the National Science Foundation's Business R&D and Innovation Survey (BRDIS) provide details of the incidence of innovation in businesses located in the United States. Later versions, however, concentrate on R&D at the expense of innovation data. In Table 4.7 we reproduce the main

Table 4.7 Companies in the United States reporting innovation activities, by industry, 2006–8 (%)

Industry	% of sample introducing new products	% of sample introducing new goods	% of sample introducing new services	% of sample introducing new processes
All	9	5	7	9
Manufacturing	22	18	10	22
Food	17	16	5	17
Beverages	17	13	6	15
Textiles	19	15	9	18
Wood	9	6	5	16
Chemicals	41	33	18	34
Plastics/Rubber	24	21	11	28
Non-metallic mineral products	13	11	5	14
Primary metals	17	13	11	19
Fabricated metal prods	16	11	9	22
Machinery	26	23	11	24
Computer/ electronic	45	43	18	33
Electrical equipment	37	36	11	28
Transport equipment	28	25	11	23
Furniture and related	14	13	6	9
Manufacture	22	17	12	23
Non-manufacturing	8	3	7	8
Information	39	16	25	20
Finance	8	1	8	8
Real estate	7	6	5	6
Professional/ scientific	13	6	12	12
Healthcare services	10	3	9	8
Other	6	3	5	7

Notes: Sample size, 1,545million.

Source: National Science Foundation, Division of Science Resource Statistics, Business R&D Innovation Survey (https://www.nsf.gov/statistics/infbrief/nsf11300/), NSF 11–300, October 2010.

sampling data related to product and process innovation in 2008. These data are based on respondents to the survey and represent an estimated 1.5 million for-profit companies, publicly or privately held, with five or more employees, active in the United States in 2008. The data cover activities in the period from 2006 to 2008. Although the pattern of the survey and of the data is very similar to that found in the European CIS survey, one should not immediately assume direct comparability.

Across all sectors, about 9 per cent of firms introduced new products, 5 per cent introducing new goods while 7 per cent introduced new services (although some businesses introduced both). Compared with the European totals detailed in Table 4.4, this is quite low and well below the leading European innovators. However, direct comparisons may not be strictly accurate. One may note that in the United States only about 9 per cent of firms were also active process innovators and thus the proportion of firms active in product and process innovation is about the same, but both figures are low in European terms.

The data indicate that the incidence of US innovation varies substantially by industry sector. For example, about 22 per cent of the manufacturing companies introduced product innovations (one or more new or significantly improved goods or services) compared to about 8 per cent of companies in the non-manufacturing sector (a pattern matched in process innovations). Both these numbers are lower than the ones found in the UK survey. In individual sectors the data indicate that the incidence of product innovation is about 45 per cent in computers/electronics, 41 per cent in chemicals (including pharmaceuticals), and 39 per cent in information, but only 8 per cent and 7 per cent in finance and real estate. Not surprisingly, the data (not presented) also indicate that companies that perform or fund R&D have a far higher incidence of innovation than do companies without any R&D activity.

4.2.3 *BRIC and other countries*

To complete the picture, in Table 4.8 we present some data on product innovation in countries outside the EU and the United Stats, including some developing countries and the BRIC economies. Again, one should warn that comparability within the sample and with other data may be limited.

It is not clear how to read these data. In some countries, for example Israel, the Philippines, Brazil, and China, product innovation is high. In other countries, for example Colombia and Egypt, it is low. In the Russian Federation it is low in general and low relative to the other BRIC economies. In US manufacturing 22 per cent of firms are product innovators. Although the figures will not be directly comparable, this might be taken as some sort of benchmark exceeded by Israel, Malaysia and the Philippines, Brazil, and China.

Table 4.8 Product innovation, manufacturing firms, various countries, 2011, % of sample

Country	% of sample
Brazil	23.0
China	25.1
Colombia	4.6
Egypt	6.0
Israel	34.2
Malaysia	29.5
Philippines	38.0
Russian Federation	8.0
South Africa	16.8
Uruguay	17.2

Source: UNESCO Institute of Statistics [UIS] (2012a), *Results of the 2011 UIS Pilot Data Collection of Innovation Statistics*. Quebec: UNESCO Institute for Statistics.

4.3 **Product launch data**

4.3.1 *Introduction*

Although CIS data are comprehensive, they have several problems that limit their usefulness: (i) they relate to the number of enterprises introducing innovations in different periods but do not provide any insight into how many innovations are introduced by those firms; (ii) except for the data on turnover arising from new products that may be indicative, the survey does not provide any information on the success of product innovations and on whether they are long-lived or short-lived; (iii) following OECD guidelines, changes of a solely aesthetic nature are not considered innovations in the CIS survey, whereas we would consider them to be so; (iv) the CIS-type surveys only look at product innovation by firms that produce in the sample countries and do not include product innovations that are imported; and (v) product innovations by newly established firms or entrants are not covered by the survey sample. To take account of these matters, we have to look at other data from elsewhere. We do this by considering a variety of data upon new product launch activity in a number of different industries and markets using a variety of different indicators. It is worthy of note that, although we concentrate on launch activity, there is good prior evidence that many new products fail (Castellion and Markham, 2013).

4.3.2 *Pharmaceuticals*

The Food and Drug Administration (FDA) in the United States has to approve all new drugs prior to their being sold on the US market. The FDA publishes

Table 4.9 Summary of number of new drug applications (United States, 1990–2011)

Year	NDAs approved	New molecular entities	NDAs received
1990	64	23	98
1991	63	30	112
1992	91	26	100
1993	70	25	99
1994	62	22	114
1995	82	28	121
1996	131	53	120
1997	121	39	128
1998	90	30	121
1999	83	35	139
2000	98	27	115
2001	66	24	98
2002	78	17	105
2003	72	21	109
2004	119	36	115
2005	80	20	116
2006	101	22	124
2007	78	18	123
2008	89	24	140
2009	90	26	146
2010	93	21	103
2011	99	30	105

Source: Summary of NDA Approvals & Receipts, 1938 to the present (http://www.fda.gov/aboutfda/whatwedo/history/productregulation/summaryofndaapprovalsreceipts1938tothepresent/default.htm). Page last updated on 18 January 2013.

on its website (www.fda.gov/AboutFD) a summary of the number of new drug application (NDA) approvals and receipts from 1938 to the present. We reproduce in Table 4.9 the relevant data on NDA for the period from 1990 to 2011.

Using the tabulated FDA data on new drug approvals, we may also look at the question of whether (original) product innovation may be becoming more common over time. Decade by decade, the average number per annum of new drug approvals is as in Table 4.10. In our view, these data do not reveal any pronounced upward or downward trend.

With around a hundred new drugs approved each year, the extent of product innovation may look quite limited. However, these new drug approvals relate only to new molecular entities, which essentially are structurally unique active ingredients that have never before been marketed. They are thus only the tip of the iceberg in terms of the number of product innovations in pharmaceuticals. Using FDA data relating to all approvals in just the month of Jan 2014 (see www.fda.gov/AboutFD), the FDA reports 186 manufacturing

Table 4.10 FDA new drug approvals by decade

Decade	Average p.a.
1960–1969	101.5
1970–1979	65.5
1980–1989	98.3
1990–1999	85.7
2000–2010	87.1

Source: The Food and Drug Administration (http://www.fda.gov/aboutfda/whatwedo/history/productregulation/summaryofndaapprovalsreceipts1938tothepresent/default.htm).

changes or additions, apart from 36 approvals of labelling revisions. This suggests much more product innovation than is apparent from the number of approved new molecular entities.

4.3.3 *Laundry detergents*

CIS data only indicate whether firms are product-innovating and not the extent to which they are innovating or how many innovations they have put on the market, although the data on proportions of turnover arising from product innovations could be indicative. To illustrate the multiple launch of new products by firms, we explored some data related to new product launches in the United Kingdom by two firms in the laundry detergents market (these data are sourced from the report 'Laundry Detergents and Fabric Conditioners, UK, August 2013' on Mintel's GNPD database). The data show that two firms, Unilever and Procter & Gamble, launched twelve and thirteen new products respectively into the laundry detergents market in a twelve-month period. Of these twenty-five new products we see that only one is listed as a new product, five are new formulations, seven are new variety/range extensions, and twelve represent new packaging. In other words, of the many product innovations launched, few are actually particularly radical. Similar data related to Pet Food and Supplies (http://academic.mintel.com, accessed on 25 March 2013) detail examples of new product launches undertaken by six companies in 2011: Butchers Pet Care (5), Mars Pet Care (6), Nestle Purina (4), Procter and Gamble (1), Armitage (2) and Bob Martin (2). Of the twenty product innovations listed, nine are described as new products, nine are labelled new variety/range extensions, one is listed as new packaging, and one as a relaunch. There is again little evidence that radical product innovations are particularly widespread.[1]

[1] In case one thinks that 'new product' means a radically new product, here is an example of the description of a new product: a tinned dog food that comprises lamb and peas, beef and carrots, and chicken and vegetable varieties.

4.3.4 *The food industry*

A food industry is found in all economies and is probably one of the oldest, if not the oldest, industries in the world. A significant postwar trend has been towards increased globalization and more extensive trade. Food has both functional and aesthetic characteristics. In developed economies such as the United Kingdom, food is no longer considered to be just a means to avoid dying or to stave off hunger pangs; it is a product that is to be enjoyed for its taste, appearance, smell, and so on. To a large degree new food products will thus reflect changes in form rather than in function.

The literature on new products in the food industry is informative. According to McNamara et al. (2003), thousands of new products are launched every year. In Germany, for example, 32,478 new products were introduced on the food market in the year 2000 (Madakom, 2001, cited in McNamara et al., 2003). Winger and Wall (2006) cite that in the United States about 18,000 new products are offered to the supermarkets each year (typically in Australia and New Zealand there are between 5,000 and 10,000) and about 10 per cent are chosen to be displayed on the shelves. New introductions to the shelves are almost always linked to the discontinuation of another product. The literature also suggests that most product innovations are failures. For example, McNamara et al. (2003) states that, of the many new products introduced on the German food market in 2000, a large share did not survive beyond the first year. Winger and Wall (2006) quote findings from the literature that show failure rates ranging from 48 per cent to 99 per cent. Feigl and Menrad (2008), using survey data, are a bit more optimistic, although they find some significant differences across countries. The picture that emerges is that large numbers of new products are being offered to the market each year, but many of these innovations do not succeed and quickly disappear from the market.

Some further information may be gained by exploring trademark data. Although data on IPR registrations will not necessarily reflect new products that reach the market (e.g. processes may also be patented, and not all patents are followed through to new products), trademark registrations are more likely to reflect product innovations; they also take place when the product is closer to market. We consider data for applications made under the Madrid system. The Madrid system is a mechanism by which an applicant may apply for trademark registration in multiple countries by filing a single international application via a national or regional intellectual property (IP) office. It simplifies the process of multinational trademark registration by eliminating the need to file a separate application in each jurisdiction in which protection is sought. In our view such data will weed out many of the unimportant applications. WIPO (2013) estimates indicate that such international trademark applications in the food and drink industries (NICE classes 29–33 inclusive) encompass 9.2 per cent of all applications made under the Madrid system.

Table 4.11 Total initial trademark registrations, NICE class 29

Year	Number
2006	2484
2007	2472
2008	2484
2009	2337
2010	2277
2011	2441
2012	2402
2013	2548

Source: WIPO, *Madrid Yearly Review*, International Registrations of Marks, Economics and Statistics series, Geneva (www.wipo.int/ipstats, various years).

Also, of the total number of trademarks across all industrial sectors registered in 2005, those in the food industry represented 10 per cent, 3 per cent, 17 per cent, and 8 per cent respectively for the United Kingdom, the United States, Germany, and Korea. This suggests that the food industry is relatively product-innovative in a wide variety of countries. The WIPO counts (see http://ipstats.wipo.org) of new trademark applications in NICE class 29 further indicates that in terms of overall registrations Europe and China lead, China showing massive growth between 2008 and 2011, whereas Europe shows a decline. One can also observe that there is much trademark registration by non-residents, which indicates that product innovation is an international activity and product innovations can be imported.

To reflect changes over time, WIPO data on trademark registrations under the Madrid system, 2006–13, in NICE class 29 are presented in Table 4.11. These data do not show any particular direction of growth.

When one looks at trademark registrations for all classes of goods and services over the period 1970–2009 (see Table 4.12), growth over time is very noticeable (registrations in 2012 were 41,954 and in 2013 were 44,414, so growth was continuing). Of course one cannot separate out increases in the amount of product innovation from increased use of the system, but a suggestion of increased product innovation over time is not ruled out.

4.3.5 *The film industry*

New films may not initially be considered important new products, but film production can require large investments. The estimated cost of making the most expensive film ever (*Pirates of the Caribbean: On Stranger Tides*, 2011) was $350 million in 2014 prices (en.wikipedia.org/wiki/List_of_most_expensive_films; accessed 1 June 2016). Such sums match those expended in the search for new pharmaceutical entities and illustrate that some innovations in the

Table 4.12 Trademark registrations and renewals, all classes, 1970–2009

Year	Registrations	Renewals	Totals
1970	10,731	2,328	13,059
1975	7,203	3,190	10,393
1980	8,028	4,310	12,338
1985	8,961	4,735	13,696
1990	17,157	4,854	22,011
1995	18,852	3,808	22,660
1996	18,485	4,510	22,995
1997	19,070	4,874	23,944
1998	20,020	5,750	25,770
1999	20,072	5,710	25,782
2000	22,968	6,869	29,837
2001	23,985	6,503	30,488
2002	22,236	6,023	28,259
2003	21,847	6,637	28,484
2004	23,379	7,345	30,724
2005	33,169	7,496	40,665
2006	37,224	15,205	52,429
2007	38,471	17,478	55,949
2008	40,985	19,472	60,457
2009	35,925	19,234	55,159

Source: WIPO (2009), *WIPO Gazette of International Marks*, Statistical Supplement for 2009, Geneva.

creative industries involve levels of investment that match those in the more often studied technological innovations.

Here we largely measure product innovation by the number of new films that are produced and marketed. In fact, as in publishing and music recording, for example, it may well be the case that, if new products are not produced by the industry, demand will fall continuously over time. One does not wish to see the same film again and again. In other words, product innovation is the sine qua non of the industry. The film industry operates in an international marketplace, characterized by a cross-border flow of products and skills. It is thus a good example of product innovation that is often sourced outside the domestic economy. It also shows that different countries may be producing their own product innovations. In addition, a count of the new films also encompasses films being made by new entrants to the industry, which other survey-based product innovation measures may not do. The industry provides insight into the successes of new products and their expected lifetimes.

Currently, major technological changes are altering the patterns of both production and consumption in the film sector, such that today most films are seen away from traditional cinemas and increasingly involve formats other than celluloid for shooting, editing, and production. Cinema admissions are an indicator of the demand for film. India had the largest number of admissions in 2009, with 2,917 million, more than double the United States' admissions (1,415 million), which in turn were more than five times those of the next country, China (264 million). In 2009 the frequency of admission in the United States was 5.2 films per year, and India's attendance frequency was 2.7. Nigeria does not show up in these statistics, although it is a major film producer. In Nigeria consumption typically does not take place in theatres but instead in what is conventionally called 'alternative screening venues', such as video theatres or semiprivate communal television sets (UNESCO Institute for Statistics, 2012). The pattern of world film production, 2005–9, is displayed in Table 4.13.

Although there is no obvious criterion by which one may judge whether the number of films produced is large or small, the overall growth rate, which averages 6.5 per cent per annum, represents an increasing rate of innovation over time. We may also observe that the leading producers of new films, India and Nigeria, are not the usual candidates for heading tables of innovativeness, although the movement of China up the league table from fifth to fourth is in line with its growing activity as an innovator in many other fields.

Further data on language (although some were missing for key countries) show that in 2009 English was by far the dominant language in film production and was used in one quarter of all films. Yoruba (11 per cent), used in Nigerian productions, was the second most common language, followed by Spanish (8 per cent), French (6 per cent), Russian (5 per cent), and Hindi (5 per cent). These six languages together represent the majority (60 per cent) of all films made in 2009.

However, of all these many films that were produced, very few are of great economic relevance. This relevance may be measured by gross revenues earned by films. Although this is an extreme example, it is estimated that

Table 4.13 World production of feature films, number of films made, 2005–9

Year	2005	2006	2007	2008	2009
World production	5658	6255	7071	7083	7233
Countries covered	95	90	95	98	100
Growth rate (%)	–	10.6	13.0	0.2	2.1
Top five producers					
India	1041	1091	1146	1325	1288
Nigeria	872	1000	1559	956	987
United States	699	673	656	759	734
Japan	356	417	407	418	448
China	260	330	411	422	475

Source: UNESCO Institute for Statistics [UIS], 2012b.

the top two grossing films of all time, adjusted for inflation (en.wikipedia.org/wiki/List_of_highest-grossing films), were *Gone with the Wind*, grossing $3.4 billion (at 2014 prices) and *Avatar*, grossing $3.0 billion; but such data are not available for all new films. Alternatively, there are some data about the most popular films. The UNESCO Institute for Statistics (2013) reports the results of an exercise to measure the popularity of films. The UNESCO Institute for Statistics (UIS) requested countries to identify the ten most popular feature films over three years, from 2007 to 2009. Most countries measured this popularity by cinema admissions, though a few used gross box-office revenue. Scoring a first-ranked film as a 10 and a tenth-ranked film as a 1, one is enabled to produce lists of the world's most popular films.

The data indicate that the popularity of films declines rapidly as one moves down the ranking of films. In 2009, whereas the most popular film merited a score of 450, the tenth-ranked scored only 57 and the twentieth only 18. Thus very few films launched are of particular market significance. It is also apparent that the top ten most popular films in 2007 and 2008 were no longer in the most popular rankings during the following year. The new products that dominate the market in one year are thus superseded by later new products in the following year.

Of the top ten films in 2007, 2008, and 2009 all were in English and either sole US productions or joint productions employing US producers. It is, however, argued by the UNESCO Institute of Statistics (UIS, 2013) that these highly visible, big-budget, English-language, franchise feature film represent only one stratum or tier of popular global film culture. This top tier, produced and promoted by the largest multinational corporations, represents an international standard for the exclusive and selective environment of the movie theatre. But, although these may be the most visible and most widely shared, there is also an active sphere that is more localized, nationally specific, in diverse languages, and likely to be enjoyed in venues other than the traditional cinema house. Locally made films may well be significant locally, but they are not globally significant (an issue further addressed in UIS, 2013).

The film industry illustrates certain characteristics that seem to be common to a number of creative industries. The first is the large number of new products that are launched each year. This indicates a very innovative sector. However, of these launches only a very few achieve large sales, and this may vary by culture or country. Some products may be locally important but not globally. Market-leading products turn over very quickly and generally have short lives, being replaced quickly by new market-leading products.

4.3.6 *Book publishing*

Individual books may represent big innovative strides as measured by market impact. The BBC (www.bbc.co.uk/news/entertainment-arts-13889578) reported on 23 June 2011 that the Harry Potter novels have sold more than 450 million copies through Bloomsbury in Britain and through Scholastic in the United States. In the absence of full data on sales, however, when it comes to book publishing, a useful approach to measuring the extent of innovation, as with films, is through data on the number of books published, although it should be noted that books may be published in one country but written and produced in another. For example, a local publication may be a translation of a book first published elsewhere. One may be reasonably sure, however, that a book published in a country will usually be in the local language. As with film, we might expect that the books published in a country reflect the culture of that country.

In Europe (Federation of European Publishers, 2012) slightly more than half a million new books are published each year; in 2011 there was an active catalogue of 8.5 million titles. These are very high rates of product innovation. Further data on new books published each year, based upon data originally collected by UNESCO, are detailed in Table 4.14. These data make clear that there is innovation even in the smaller countries. Approximately 2,200,000 new books were published in total across the 116 countries in the UNESCO sample. The dominance of China is a recent phenomenon. The large number of new publications in the United Kingdom and in the United States implies that the most common language in use for new publications will be English.

Many new titles are irrelevant as innovations of economic significance. Only a few are significant. We use market share to indicate significance. Using data on fiction book sales from the *New York Times*, Sorensen (2007) observes that, of the 1,217 books for which sales data were observed, the top twelve (1 per cent) accounted for 25 per cent of six monthly sales and the top forty-three (3.5 per cent) accounted for 50 per cent of sales. The 205 books that made it onto the *New York Times*' bestseller lists accounted for 84 per cent of annual sales in the sample. This indicates that, to encompass the vast

Table 4.14 New books published per country per year

Rank	Country	Year	Number of titles
1	China	2011	369,523
2	United States	2010	328,259
3	United Kingdom	2011	149,800
4	Russian Federation	2012	116,888
5	India	2004	82,537
10	Pakistan	2010	45,000
20	Indonesia	2009	24,000
30	Thailand	2009	13,607

Source: Wikipedia, 'Books published per country per year' (https://en.wikipedia.org/wiki/Books_published_per_country_per_year, accessed 16 September 2013).

majority of sales in the industry and thus the most significant innovations, one does not have to go far down the sales order.

Using weekly data from 'The New York Times Best Sellers' for fiction titles from 1970 until 2004 grouped into seven five-year periods (1970–4, 1975–9, 1980–4 ... 2000–4) and calculating, for each five-year period, the number of titles that appeared in the top ten and for how many weeks these titles remained there, Stoneman (2011) shows that the number of titles that entered the top ten per period has increased significantly since 1970–4, from 166 in the first period to 491 in the last period. It is also clear that the number of weeks that a book spends in the top ten has fallen considerably between 1970 and 2004, from 15.7 to 5.3 weeks—which suggests that there has also been a significant increase, over time, in the proportion of sales that arise from recently launched product variants. The fact that in 2000–4 a title spent on average 5.3 weeks in the top ten is equivalent, on average, to 19 per cent of sales being from titles new to the top ten each week. Under either metric, the data suggest that the rate of innovation, as indicated by the number of significant new products launched, is high and had increased three times over thirty years.

Once again, therefore, we see a large number of new products being launched each year. This indicates a very innovative sector. As in film, however, only very few of these launches achieve large sales, although which ones do so may vary by culture or country (and some products may therefore be locally important but not globally). Market-leading products here too turn over very quickly and generally have short lives, being quickly replaced by new market-leading products.

4.3.7 *Recorded music*

There are now quite a few formats in which recorded music is sold (or pirated). While vinyl (although currently experiencing a renaissance) and cassettes are now almost defunct, CDs are still widespread, but electronic downloads are now supreme (www.ifpi.org/content/library/dmr2013). Data for 2015 on this industry are available in the IFPI Global Music Report 2016 (http://ifpi.org/news/IFPI-GLOBAL-MUSIC-REPORT-2016), which shows that global industry revenue was US$15.0 billion. Although the detailed statistics from this source for 2015 are only available at a cost we cannot bear, in 2012 global recorded music industry revenues were US$16.5 billion, Adele being the best selling artist with 8.3 million units sold. The United States retained the top position, its sales totalling $4.48 billion. The world's number two market remained Japan, its recorded music sales totalling $4.42 billion. In third position was the United Kingdom, where music sales totalled $1.33 billion; the series continued with Germany, $1.29 billion, France, $907 million, Australia, $507 million, Canada, $453 million, and Brazil, $257 million (these figures are sourced from www.billboard.com/biz/articles/news/digital-and-mobile/1556590/ifpi-2013-recording-industry-in-numbers-globa, accessed on 23 September 2013).

Given the problems of data access, we rely for our analysis upon some data that are freely available and relate to Japan, which is the world's number two market. The data from the Recording Industry Association of Japan (2016; see www.riaj.or.jp/e/issue/pdf/RIAJ2016E accessed on 1 June 2016), indicate that 11,725 new 5-inch CD albums were released on the Japanese market in 2015, the back catalogue of such CDs standing at 132,392. This indicates an industry where large numbers of new products are launched on the market each year, as in book publishing. However, in order to get a clearer picture it is important to attempt to separate out from the large number of titles being launched each year those that are a success in terms of market share and thus of market impact.

Using UK data, Stoneman (2011) shows that (i) only 0.7–0.8 per cent of the albums released enter the top forty sales charts at any time; and (ii) sales of any album, as a proportion of sales of the highest selling album, decline sharply as one moves down the rank of sales. It is obvious that sales soon die away as one moves down the bestseller lists. Stoneman (2011) also shows that about nineteen new album titles on average chart each month, suggesting that on average about half the top forty titles will change each month or that there will be complete churn of the top forty on average every two months. The overall picture of product innovation in the music industry is similar to that found in books and films. There is very extensive innovation, as indicated by the data on launch patterns, but many titles fail and only a few are significant in terms of sales. There is also a high rate of churn in the top titles.

Stoneman (2011) explores the video games industry, which is shown to launch numerous new titles each year as well. Few, however, succeed. The life cycle of a new successful title is short (the data suggesting that in a year at least thirty-six titles will have been in the top ten, which is close to the figures calculated for books), and there is evidence that life cycles are getting even shorter, as games seem to hold their ranking (e.g. at number one) for increasingly shorter periods.

4.4 **Scanner data**

In the separate parts of section 4.3 we have explored product launch data for a number of different industrial sectors. An alternative source to explore is that of scanner databases generated by stores from the bar codes of the goods they sell. Although this kind of source has some appeal, access to such data is often limited. Perhaps equally informative (if not even more) is the related work of Broda and Weinstein (2010), who analyse the nature and extent of product entry and exit in US consumer markets using barcode data derived from the ACNielsen Homescan database. ACNielsen provides handheld scanners to approximately 55,000 households, whose members then scan in their purchases of every good with a barcode. The database represents a demographically balanced sample of households in twenty-three cities in the United States with the barcodes concentrated on grocery, drugstore, and mass merchandise sectors. Overall, the database covers around 40 per cent of all expenditure on goods in the consumer price index (CPI).

The data include the average price paid and the total quantities of all products with universal product codes (UPCs, also referred to as barcodes) purchased by the representative household at the quarterly level for six years: 1994 and 1999–2003. The data were weighted by ACNielsen to correct for sampling error. There are three levels of aggregation. At the lowest level are the UPCs. At the intermediate level, each UPC in turn belongs to a brand module; and at the highest level there is the product group. Roughly 650,000 different UPCs are identified as having been sold in the sample period. In the data, the average firm sells forty UPCs under four different brands in three modules, which are in turn contained in two product groups. However, in the fourth quarter of 2003, 60 per cent of the sales came from firms that sell over 700 UPCs in over thirty-five different brands and nineteen different product groups, that is, the bulk of sales in these sectors is on goods produced by firms marketing hundreds of different products under dozens of brands in a variety of markets.

Broda and Weinstein (2010) report a number of results based on these data that give us some insight into the rates of product innovation and destruction in markets. First, indicating most directly the rate of launch of new products, the data show that in a one-year period the median entry rate (number of new products in time t/total number of products in time t) was 0.25, that is, in a given year new products represented 25 per cent of all products on the market, whereas the median exit rate was 24 per cent. Over a four-year period (1999–2003) the rates were 50 per cent and 46 per cent respectively, whereas over a nine-year period (1994–2003) the entry and exit rates were 78 per cent and 72 per cent respectively. These numbers suggest that in the markets under consideration the rate of new product entry (and exit) was high by any relevant yardstick, and hence one must consider that product innovation is a very common phenomenon.

It is also found in the data that net creation is strongly pro-cyclical, more products being introduced in expansions and in product categories that are booming. Destruction of goods is countercyclical, although its magnitude is quantitatively less important. Of course not all product innovations are a success. Taking the extent of product sales into account, in a typical year, 40 per cent of household expenditures was on goods that were created in the last four years and 20 per cent of expenditures were on goods that disappeared in the next four years. These same data imply that 37 per cent of the household expenditures in 2003 were on goods that did not exist in 1999, and 18 per cent of expenditures in 1999 were on goods that did not survive until 2003. Alternatively, almost 80 per cent of the products that existed in 2003 were not around in 1994, expenditure on these new products comprising 64 per cent of expenditures in 2003.

The market share of the average new UPCs is 30 per cent, which is as large as the share of common UPCs. For the average exiting UPCs, this share is 9 per cent. This suggests that new UPCs tend to be larger than existing ones, but both are smaller than common UPCs by a large margin. In addition, the value of the UPCs that existed in 1994 but did not exist in 2003 was only 37 per cent of the expenditure in 1994. This suggests that the new UPCs systematically displaced market share from the UPCs that were common throughout this period. This can also be summarized in terms of the ratio of the shares of common UPCs that was below unity; in other words the share of common UPCs in 1994 was larger than in 2003. Broda and Weinstein (2010) consider this to be an important indication that new products are of higher quality than the products that exited before.

By matching barcodes with firms, the data also illustrate the multi-product nature of firms and shows that most product creation and destruction happens within the boundaries of the firm. In particular, it is found that there is four times more entry and exit in product markets than in establishment and labour market data. Broda and Weinstein (2010) compare the extent of firm creation to that of product creation, which suggests that in a one-year period 92 per cent of product creation happens within existing firms, and 97 per cent of product destruction happens within existing firms. At four-year frequencies the comparable numbers are 82 per cent and 87 per cent respectively. This implies that, over a four-year period, 18 per cent of the value of overall consumption is coming from products of completely new firms, and 13 per cent of product exit is happening because firms disappear. Product entry and exit are five times as important as firm entry and exit.

A further aspect of the data concerns how entry and exit co-vary across product modules. Using one-year entry and exit variables indicates a correlation between entry and exit variables across modules that is positive and high, the average contemporaneous correlation between entry and exit rates being 0.70. In terms of creation and destruction (where creation is measured as the ratio of sales of new products/total sales and destruction is measured as the ratio of the sales of disappeared products/total value), the contemporaneous correlation is slightly higher at 0.83. This suggests that modules tend to be characterized by either high entry and exit or low entry and exit. Thus destruction rates are higher for smaller and younger UPCs. Destruction does not monotonically fall with the size of the brand. Small and large brands have higher rates of destruction than middle-sized brands.

4.5 **Industry maturity and product innovation**

We have illustrated that in many quite different industries product innovation is widespread, a large proportion of firms undertaking such activities and in many cases introducing many new products on the market in a given period of time. There is a stream of literature, however, that suggests that to look at snapshots of industries at a point in time is rather misleading in that industries pass through a 'life cycle' and that during this life cycle the tendency of firms to product innovate varies (see, for example, Geroski, 2003).

This life-cycle approach is often associated with Abernathy and Utterback (1978; see also Utterback, 1994), who identify three phases in industry development: fluid, transitional, and specific. At the beginning of the fluid phase, an original product appears on the market. This is the birth of the industry. After the introduction of this new product, competitors enter the market and start to sell comparable goods. New entry means, of course, product innovation in the industry. However, at this stage there is no agreed or dominant design in the industry (often defined as an industry or product standard in the economics literature) and early participants in new industries are free to experiment with new forms and materials, unencumbered by universal technical standards or by uniform product expectations on the part of buyers in the marketplace. As firms test markets and technologies, many and varied new products will be placed on the market. Many of these innovations will of course fail. Many may not be compatible with other products on the market. At this stage, too, the production process tends to not be very well developed (Utterback and Abernathy, 1975). Often quoted examples of this path to a dominant design are (i) the emergence of personal computers in the 1970s and the dominant IBM-compatible design; (ii) the development of home video and the eventual dominance of VHS; and (iii) the competition between blu-ray and HD-DVD standards. Once a dominant design has been established this stage of intense product innovation is over.

In the transitional phase the rate of intense product innovation is over and standard designs are in place. As the market now has particular expectations for a product, the bases on which product innovation can take place become much fewer and the practicalities of marketing, distribution, maintenance, and so forth may generate even greater standardization. Process innovation becomes the major innovative activity. Although product innovation still occurs at a lower level, the main competition is via prices, that is, via cost reductions.

In the final specific stage the rate of both major product and process innovations declines, industries becoming mainly focused on cost, volume, and capacity, as products are standardized and undifferentiated. The probability of new radical innovations decreases due to the standardization processes and, as a result, both product and process innovation rates decrease (Milling and Stumpfe, 2000). The focus is mainly on efficient production and cost minimizing in order to be able to sell at the lowest price possible (Utterback and Abernathy, 1975).

It has been argued that the model of Utterback and Abernathy (1975) is typical of many industries such automobiles, typewriters, bicycles, sewing machines, TV, and semiconductors. However, although the approach of Utterback and Abernathy (1975) lays particular emphasis on the rate of product innovation and thus merits our attention, it should be noted that, for different industries, the different phases may take (considerably) different periods of time. Thus one cannot say that after n years an industry will be moving from phase 1 to phase 2 and hence expect the rate of product innovation to slow. All one can do is suggest that industries that are in the early phase may show more product innovations than those at a later stage. This does cause one to be wary of inferring causality. Nor is it the case that the model will necessarily apply to all industries. We have suggested that the creative industries have illustrated very high rates of product innovation for a considerable period of time. In fact we have suggested that such innovation is a sine qua non of some industries. The upshot of this is to argue that some industries will always be in the first of the three phases. Finally, the Utterback and Abernathy (1975) approach is not the only approach that attempts to look for industry life cycles. Sutton (1998), for example, explores the development patterns of industries (which also reflect a similar time profile, e.g. the aircraft sector), indicating the main causal factors over time.

One issue of particular relevance is the rate of entry and exit to industries, for the more new entrants there are to an industry the more one might expect there to be greater product innovation, each new entrant offering its own new product. Klepper and Grady (1990) have explored the evolution of the number of producers in a sample of forty-six new product categories in the United States from initial commercial introduction up until 1981, the products having initial commercial introductions from 1887 on. They characterize industry development as having three stages. At stage 1 the number of firms supplying the new product grows. This stage is supposed to finish when the maximum number of producers is reached. At stage 2 there is a decline or shakeout of firms. Stage 3 is a period after the shakeout when the number of firms stabilizes. We expect most product innovation to happen at stage 1.

The data indicate that in some cases the peak number of suppliers occurs very early (e.g. Fluorescent Lamps), whereas other products show an increasing number of suppliers for a very long period (e.g. Gyroscopes and Electrocardiographs). Some industries (e.g. nylon) have not reached the second stage even after many years. At stage 1 the number of suppliers may be increasing, but the rate of increase differs across industries. The average mean annual change in the number of suppliers at stage 1 is 3.8, but in some industries it is low for a long time, and in yet others high for a short time. Thus for example in Rocket Engines the number of suppliers increases for thirty-seven years at an annual average rate of 0.4, while in Fluorescent Lamps stage 1 only lasts for two years, but in those two years the number of firms increased by thirty-two. The number of suppliers changes little at stage 3, when there is an average increase per annum of only 0.2. The extent of the shakeout at stage 2 also differs across industries, for example in Tires, out of a peak number of 275 suppliers, 211 left at stage 2, whereas in Shampoo, out of 114 suppliers at the peak, only 5 left at stage 2.

These data of Klepper and Grady (1990) show how the number of suppliers changes over time and how the actual path of evolution is different in different sectors. Although the dominant pattern is one where there is an initial phase with an increasing number of firms followed by a phase with a decreasing number, then reasonable constancy, the length of these phases differs considerably across products. Moreover some industries have never reached stage 2, not even after a long period of time (thirty-seven years in Rocket engines), whereas others reached it quickly and did not stay there long (outboard motor spent only fifteen years in stages 1 and 2 in total). There is thus considerable heterogeneity across industries, even if the general pattern observed in this sample is common. This suggests that there may well be considerable heterogeneity across industries in rates of product innovation (and death) as industries are born, mature, and decline, even if only as a result of firm entry and exit (see also Klepper, 1996).

4.6 **Conclusions**

In this chapter we have provided a number of indicators of the prevalence of product innovation in economies. We have allowed that product innovation may come from existing firms or from new entrants and may also be imported. An initial analysis of survey data on the extent of product innovation in Europe indicated that, in the three years prior to the 2012 survey, about 25 per cent of all firms in the twenty-seven EU countries undertook product innovation. We showed that some of this innovation was new to market and some new to firm, the latter being more frequent. We also showed that, in general, new-to-firm innovations contributed more to firm turnover than new-to-market ones (remember that the former is more frequent). We were able to look at differences in rates of product innovations across sectors, regions, and firm sizes. We also undertook some comparisons with similar data for the United States, where there seemed to be lower incidences although considerable differences across industrial sectors.

Although the survey data are informative, they only indicate whether firms undertook product innovation and not the extent to which they innovated. We thus proceeded to look at the extent of new product launches in a series of different industries using other data. From the laundry detergent industry we saw that the two main firms there were introducing twelve or thirteen new products per annum. However, we also observed that many of the new products were not significantly different from exiting products, involving just new packaging or brand extension rather than significant changes in characteristics.

Looking at pharmaceuticals, we also saw that, although there are many product innovations of limited importance, new chemical entities are approved each year. In our view these data do not reveal any pronounced upward or downward trend, and in consequence we do not observe any weakening in the tendency to product innovation in the industry. A study of the food industry also revealed very extensive product innovation in many different countries. This was illustrated using data on trademarks. Here we also observed for the first time that the rate of failure of product innovation may be very high, very few new products succeeding and surviving.

From the food industry we moved to analyse three creative industries: the movie industry, book publishing, and music recording. In each case the number of new products being launched annually was very high. In fact one might consider that in these industries new products are a sine qua non of firms' continued existence. One does not wish to read the same book, see the same film, or hear the same music over and over again. One wishes to experience new books, films, and music. We also showed that each country may have its own products and product innovations that suit the local culture, although there may well still be much exporting and importing of new products. Importantly, we also illustrated that, out of the very large number of new products that are put on the markets, very few succeeded. Only a few books sold in large numbers, very few films got large audiences, and very few recordings sold widely. In this sense one may say once again that most product innovations fail.

In an analysis of barcode data relating to all consumer purchases, Broda and Weinstein (2010) also provide an informative guide to the extent of product innovation (and death) in markets. They indicate that in a one-year period the median entry rate was 0.25—that is, in a given year new products represented 25 per cent of all products on the market, whereas the median exit rate was 24 per cent. Over a four-year period (1999–2003) the rates were 50 per cent and 46 per cent respectively, whereas over a nine-year period (1994–2003) the entry and exit rates were 78 per cent and 72 per cent respectively. As we already stated, these numbers suggest that in those markets the rate of new product entry and exit was high by any measure and that, in consequence, product innovation is a very common phenomenon.

Finally, we discussed the possibility of industry life cycles to which product innovation activity may be linked. We illustrated that industries move through phases and in the early phase product innovation is most prevalent. Although there is supporting evidence for this, there is no standard for the length of the first phase and the model may not apply to all industries. We also argued that firm entry and exit may also go in phases, and thus in the early phase of maximum entry one may expect more product innovation. Once again, however, the model does not apply to all industries and phases may have different lengths in different industries. Nor is it possible to say that industry life cycles are necessarily causal of product innovation—they may just be descriptive.

5 The Demand for a New Product

5.1 Introduction

This chapter is the first of three interconnected chapters that jointly attempt to review the factors that drive the demand for and supply of new products and the resultant incentives to undertake product innovation. This first chapter explores the determination of the demand for newly launched products, with a particular emphasis upon intertemporal development of demand. The following chapter discusses the pricing and output decisions of the suppliers of new products, whereas the third considers the decision to launch new products.

Our emphasis in these three chapters is upon products that are produced in significant numbers. These may be goods, for example mobile phones, or services, for example insurance products. They may be in the creative sector for example books, theatrical productions, movies, or in other sectors of the economy, for example rubber products. They may be producer goods or consumer goods. To reflect such heterogeneity, we will try to address all these types of products as the chapters proceed, even if only to point out where such distinctions matter. What we will not be addressing, however, is products that are specially commissioned for example architectural services, customized products, products that are unique, such as original works of art, or products produced in very limited quantities. In such cases many of the issues that we address may still be relevant but operate in rather different ways.

In this chapter much of the literature that we consider has been produced within the framework of the study of the diffusion of new technology, and thus what we say here may overlap with works in that field (Stoneman, 2002, Stoneman, 1989, and Stoneman 1990; the latter two explicitly consider the diffusion of new products that are respectively vertically and horizontally differentiated from existing products). It should be noted that here our main emphasis is upon the demand for the new product rather than on the possible market outcome, taking into account supply factors. Our approach may also overlap with that literature which considers the product life cycle. We do not, however, place much emphasis on that area.

5.2 The base case: an original, producer durable, monopoly supplier

We start by considering a new product that is a producer durable, is original in that no similar product is available on the market, and is supplied by a monopolist. One might think of an innovative machine tools as an example. It is immediately clear that demand for the new product at a point in time will be related to the extent of knowledge about the existence of the product. The more potential buyers who know about the product, the greater the demand for the product will be, *ceteris paribus*. One may also reasonably argue that the decision of a potential buyer as to whether to demand such a product will be positively related to the annual gross profit gain that may be generated by use of the new product. This annual gross profit gain will in turn be related to the prices at which the output of the producer can be sold and to the cost of other inputs. There may also be costs of training and possibly other organizational adjustments necessitated by the new product. For example, the new product may require different skills in the workforce, or a different workforce, and there is considerable evidence that firms rarely just undertake product innovation or process innovation alone but generally undertake organizational and managerial innovation as well, especially those firms that are most successful (Battisti and Stoneman, 2010). Further determinants of the net return to the firm from introducing the new product are the rate of discount and the cost of purchase of the new product. The gross and net profit gains to be derived from the new product may either not be known to the firm or may be known only with some uncertainty. The decision to make demand for the new product effective is thus one that involves risk and or uncertainty; in consequence, one might expect that the greater the risk involved the fewer the buyers are likely to be, *ceteris paribus*.

Thus buyer information, product performance, adjustment costs, riskiness, prices, and discount rates may all impact upon the demand for a new product at a point in time (among other factors). Of more interest, however, is how things may change over time. We begin by considering learning and the extent of knowledge about the existence of the product. There is a long tradition in the study of the diffusion of new products that argues that over time knowledge about the existence of a new product is self-propagating (Mansfield, 1968). Essentially it is argued that early users demonstrate the existence of a new technology to those not previously informed, and thus the extent of knowledge of its existence increases with use and with time. It is often further argued that, as time goes on, the number of non-informed potential users decreases, while (assuming that knowledge of existence leads to further usage) the probability of their observing a user over time increases, and jointly the two effects produce a time path of knowledge and usage that is S-shaped (or sigmoid), starting at a low level, increasing quickly up to a point of inflection, and then increasing at a decreasing rate to an asymptote.

Most literature assumes that the spreading of knowledge of existence is a purely domestic affair. However, knowledge flows may well be global rather than domestic and as such use overseas may impact upon knowledge at home and thereby impact upon the intertemporal pattern of domestic demand (Pulkki-Brännström and Stoneman, 2013). Of course self-propagation may not be the only means by which knowledge of existence spreads. The supplier may undertake advertising in order to inform potential buyers of a new product, for example by advertising in trade magazines or through displays at conferences. Alternatively government policies may attempt to trumpet the advantages of new technologies through demonstration activities.

However, knowledge of existence would not of itself be sufficient to persuade potential buyers to express an effective demand for a new product. Knowledge of performance of that product is also of importance. Once again, there is a tradition of modelling the spread of such information flows as the result of self-propagation. This starts with the work of Mansfield, who argues that observation of early uses informs potential later users about the profitability of a new product. However, in Mansfield's (1968) approach, the process is one in which greater knowledge only reduces the risk associated with using a new product, the potential profit being considered unchanging. In Stoneman (1981) this is modified to allow for Bayesian learning, so that user learning from observation changes not only the risk (variance) associated with the expected return but also the expected return itself.

An alternative form of self-propagation may be considered when the new product is divisible, or alternatively where the buyer may have the potential to introduce a number of units. In such cases the buyer may start by introducing a limited amount of the technology and then, by observing the performance of that small amount, s/he gathers information that informs his/her later decisions. In such cases the learning potential is in fact a reason for acquisition, and it might even be that a firm would acquire a small amount of the new product in order to learn about the product, even if it proves to not be profitable. One may observe in fact, that over time, more and more firms demand the new product (this is increased inter-firm diffusion), but the intensity of that demand by each firm (intra-firm diffusion), although still increasing, lags behind (see Pulkki-Brännström and Stoneman, 2013).

It is also possible that a firm acquires information via active rather than passive search (Jensen, 1988). Thus the firm may undertake research activities related to the potential of a new product that encompass for example internal research, external research, discussions with providers, and close observation of existing users. It is important to note, however, that, because

use by others (known as the stock effect) may reduce the benefit to a user, users may have an incentive to retain their acquired knowledge internally rather than inform other potential users either explicitly or through behaviour (Stoneman, 2013).

How buyers will make decisions when faced with imperfect information, that is, with uncertainty, is still an open question. One approach is to argue that buyers trade off risk and return, and perhaps decide by portfolio adjustment (Markowitz, 1952) whether or not to invest in a new product at a point in time. The real-options approach (Dixit and Pindyck, 1994), however, argues that potential buyers will delay adoption and seek to obtain further information over time until they have sufficient knowledge to make further waiting not worthwhile, whereupon adoption occurs.

It is not, however, necessarily the case that self-propagation arises solely via the spreading of knowledge. There is also the potential for network effects (Stoneman, 2013). Basically network effects arise when increased usage impacts upon the return to either all or at least the marginal adopter. In most tellings of this story the impact is considered positive, so that increased usage, *ceteris paribus*, increases returns to potential buyers. Such network effects for the base case discussed here include, for example, a growing supply of specialized labour, an increased availability of joint inputs to the production process, or greater standardization in product markets. Such effects may well impact (inter alia) upon adjustment costs, encouraging adoption.

In contrast, there is also an argument that potential first-mover or early-mover effects may impact upon the demand for a new product. In such circumstances the first or early buyers have the potential to get a greater return from use of the new product than later users or purchasers (for an early discussion, see Van der Werf and Mahon, 1997). Lieberman and Montgomery (1988) summarize the reasons for the existence of early-mover advantages, which incorporate proprietary learning effects, patents, pre-emption of input factors and locations, and development of buyer switching costs. Conversely, they also argue that early-mover disadvantages may result from free-rider problems, delayed resolution of uncertainty, shifts in technology or customer needs, and various types of organizational inertia. In addition, early-mover advantages may arise as an early mover may have the potential, via the adoption decision, to affect the intra-firm diffusion decisions of later users—a pre-emption effect often attributed to Fudenberg and Tirole (1985). Although early movers may also face some disadvantages (e.g. greater uncertainty), examples where positive benefits may arise are that early buyers may acquire the best locations or market positions or may appropriate the best inputs. In an early study, Lambkin (1988) shows that pioneers tend, on average, to outperform later entrants but, by comparison to early followers and late entrants, they have significantly different strategic profiles and performance levels. Schnaars (1994) on the other hand argues that being an imitator is the best route. In a more recent study, Lanzolla et al. (2010) show that in the European mobile telephony market, over the period 1998 to 2007, first and second movers consistently outperformed later entrants. If there are early-mover advantages, then the effect will be to shift demand for a product earlier in time; but, more importantly, the effect will be to reduce the demand for the new product by non-adopters as the number of prior users increases—the later users being unable to attain the higher returns gained by the early users and perhaps being deterred by the actions of earlier users.

Let us turn, then, to the relationship between demand for the new product and its price. In the simplest case one may consider that, *ceteris paribus*, the lower the price of the new product, the greater the demand for the product will be. This effect may arise in two ways. First, the number of units demanded by a single buyer may increase as the price of a unit falls. Thus, for example, demand for laptops by a producer may increase as the price of laptops is lowered. Secondly, price may impact upon the number of buyers rather than upon the number of units acquired by each buyer. For example, potential buyers may have different characteristics (some are large and some are small) and thus be able to derive different gross returns from a new product (rank effects). In such circumstances, firms with certain characteristics (say, large) are willing to pay a higher price for the new product than firms with other characteristics (say, small), and thus, as the price reduces, more (say, smaller) firms will demand the new product. The actual characteristics that matter may be many. Apart from size one might consider complementary inputs used, knowledge bases, location, age of existing equipment, and so on.

Although this simple effect may be most apparent, there is also the possibility that the expectation of reductions in price in the future will reduce the demand for a new product today. If it is expected that a durable product will be acquired more cheaply tomorrow than today, although the extra return today may be sacrificed, if the expected decrease is large enough (and the discount rate not too small), a potential buyer may well find it worth waiting until the later date before demanding the new product. There is the potential for these price expectation effects to interact with (counteract) early mover advantages.

Implicitly, we have assumed thus far that the potential buyers in this base case are producers in an industry with a number of other producers. Interactions between the producers are the basis of a number of the models of learning, risk reduction, first-mover advantages, and so on. More directly, however, competition between producers on the product's market may feed back into the demand for the new product that is being considered. Basically it is argued that, as more firms acquire the new technology, costs fall and, as costs fall, the market price the firms receive for their output will fall. This reduction in market price will impact upon the gain to be made by the marginal adopter. It is generally argued that this gain will fall as the market price falls, in other words greater usage of the new technology reduces the benefit to be obtained by the marginal adopter and thus reduces further demand for the new product (Reinganum, 1981). This effect is known as a stock effect and its size may well be related to, for example, the degree of competition among potential buyers and firm size.

As we are here considering a new product that is a producer durable, the stock effect, the pre-emption effect, and the rank effects among others imply that, as time goes on and more of the potential buyers acquire the product and perhaps acquire it in increasing amounts, the number of potential new buyers will decline, that is, today's additional demand for a new product may be inversely related to the number acquired to date. Although we have argued that price reductions may increase demand for a new product, these arguments suggest that, in the absence of price reductions, demand may eventually fall to zero, that is, the market might become saturated at the current price. Although there may be a possibility of replacement demand if the durable product has limited life, essentially demand for the new product would eventually tend to zero if its price stops falling. Of course this situation would be alleviated for the supplier of the new product if some improved or different variety of the product were to be put on the market.

5.3 Consumer durables and non-durables

Thus far we have considered a new product that is original and durable and purchased by producers rather than by consumers. The purchasers were assumed (implicitly and explicitly) to be seeking profit gains or, in a more intertemporal context, to be concerned with the market value of the firm. We now turn to the case where the product is a consumer good or service and hence the potential buyers are assumed to seek gains in utility rather than gains in profits. This change does not, however, introduce great change in to the potential determinants of the demand for a new product. For example, heterogeneous buyers and price expectation effects may still be important; nevertheless, some of the effects detailed for producer goods may be of different importance in the case of consumers. One might argue that first-mover or early-mover advantages may not be large for consumers. Although it has been argued that via Veblen effects (Veblen, 1899) early adopters may gain some extra kudos, the main early-mover advantages arise from gains derived via the ability of the early mover to affect the decisions of others. Such gains would seem small in the consumer's case. Similarly, it is not easy to envisage why stock effects (ignoring network effects) may be particularly important to consumers.

In the consumer case, learning effects and issues about risk and uncertainty may continue to be relevant. However, consumers may not be large spenders on research. In many cases they will have neither the resources nor the incentive to undertake costly research, given the proportion of income that is likely to be spent on the innovative product. This does not indicate that they will not search, but rather that they may make use of secondary sources, such as consumer guides and internet sources, to inform their decisions. Of particular relevance, however, are consumer goods that require some additions to knowledge before they can be successfully used. Thus home computers, smart phones, many electrical products, even vehicles require some learning and training before they can be used to their fullest extent. Such capability acquisition may well impact upon the demand for products. As usage becomes easier over time, one might expect demand to increase.

Network effects may also be important in the consumer case. Thus, for example, the growth and improvement of the road network and the availability of petrol stations would have impacted significantly on vehicle ownership by households. Similarly, in the early days of consumer video technology, the availability of hired films on video increased the demand for video players.

Perhaps the consumer case is of more interest for the consideration of a product innovation that is non-durable rather than durable. A non-durable, if permanently adopted, will involve repeat purchase, whereas a durable is more likely to involve a one-time acquisition. Non-durables may be goods (e.g. motor oil) or services (e.g. banking services). In some cases it is possible to consume a product as both a non-durable and a durable. Thus one may attend the cinema to see a film. This is purchase of a non-durable. Later, however, one may purchase a video of the film; this would be acquisition of a durable product.

A major difference between a durable and a non-durable is that decisions about non-durables are reversible, in other words the purchase does not commit one to ownership in the future. This suggests that real-options decision-making would no longer be relevant. Moreover, the trade-off between risk and return may be far less crucial for non-durables. In fact the demand curve for a non-durable product innovation may be very similar to that for an established non-durable product, that is, demand would be a function of price: demand would increase as price falls, either with each purchaser buying a greater amount or with purchasers of different characteristics becoming new consumers. The demand curve would, as in standard textbooks, shift to the right as incomes increase. The demand for a repeat purchase product is also unlikely to reach saturation, where demand falls to zero at all prices above marginal costs.

The demand curve for a product innovation may, however, differ from that for an established product, in that there may still be issues related to information spreading to consider. An original product innovation may be unfamiliar to the market. As information about the existence and performance of the product innovation spreads, one might expect (if the product is superior) that demand for the product will increase. Information may spread in a self-propagating way, via those who have purchased the product in the past (at home or overseas), or in more direct ways, via advertising by suppliers (Glaister, 1974). In addition, if the product is non-durable, information may be gathered by consumers 'buying to try', with the possibility, of course, that consumers will try and then not buy again.

5.4 Innovations new to the market

Thus far we have considered demand for an original product with a monopoly supplier. Over time, however, other new-to-market products will be introduced, either by the same or by other suppliers. In this section we explore resulting issues related to the demand for (i) the new-to-the market product(s); (ii) the original product offering; and (iii) the overall demand for the product class.

A crucial distinction to consider when exploring the impact of the launch of new-to-market products on demand is whether, compared to the original product, the new-to-the-market product is horizontally or vertically differentiated. For two product variants, when, at a common price, there is no consensus on preference ranking across potential consumers, some preferring one and some preferring the other, the two products are considered horizontally differentiated (Hotelling, 1929). When there is a ranking, all buyers at a common price preferring the same product, the products are called vertically differentiated (Gabszewicz and Thisse, 1979, Shaked and Sutton, 1982).

Let us consider first the case where the new-to-market offering is vertically differentiated from the original product. In this case, the new variant has performance characteristics that differ from those of the original product. Most of the literature considers that the new variant has improved performance. Thus, for example, later generations of computers, intelligent phones, and so on all offered improved performance relative to the original. Although there may be some issues related to learning about the new variant and issues of brand loyalty, in essence, if the new-to-market product has the same price as the original, by offering an improved price performance ratio, one might expect the new-to-the-market product to take the market from the original product (even if not overnight) and also to increase the demand for the new product in total. The original product will be superseded.

If the price of the new variant is above that of the original product, then it may only capture some of the market rather than all of it. Buyers who particularly value the dimensions improved on the new-to-market product will either be persuaded to buy the product class for the first time or, if they are repeat purchasers, to choose the new-to-market product rather than the original. Overall, the increase in total demand for the two variants may be less than if the price of the new product was lower. If the price of the new-to-market variant is lower than that of the original, one may expect that not only will the original product more quickly disappear from the market, but also joint demand for the two variants will increase more than if the price of the new variant is the same as the original. One may of course note that, in reaction to the new-to-market product, the price of the original may be reduced, thus bolstering the demand for the original, reducing the demand for the new-to-market variant, but increasing the joint demand.

As stated, the usual case is where the new-to-market product offers better performance than the original. But this need not always be so. A new-to-market product may be agreed by all to offer less than the original but, by being very cheap, still offers a better ratio of performance to price. This is common in the electronics world, where less powerful machines or machines with fewer or reduced levels of characteristics (e.g. memory) are offered on the market and succeed because they are sold at much lower prices. One may consider as well how a motor-car range offers different levels of trim and power at different prices, often very basic models being offered at very low prices. One might well expect that the improved ratio of performance to price will take the market away from the original product (partly or totally) and also increase demand for the product class as a whole.

Instead of being vertically differentiated from the original product, a new-to-the-market product may be horizontally differentiated. In this case there is no universal agreement that the new product is better than the original. For some buyers that is so, for others it is not. If the new-to-market product has the same price as the original, some of the buyers of the original product may shift to the new product. However, other potential purchasers who were not sufficiently appreciative of the characteristics of the original product may prefer the characteristics of the new product and become purchasers for the first time. For example, a soluble form of a medicine may be taken by some who would not have used the pill-based version. The horizontally differentiated product thus has the potential (if priced the same as the original) to both take market from the original and increase the overall market demand. As the price of the new-to-market product is relatively lowered, the two effects will be more pronounced. This illustrates that a new-to-market product does not have to be better in order to attract demand and increase product sales. In some cases it only has to be different.

Even with this argument, there are some aspects that may require caution. A particular example of variants being different is that they may belong to different networks or be based on different operating standards. A famous example was that of videotape recorders where there were three de facto standards: VHS, Betamax, and Phillips Video 2000 (see Cusumano et. al.,

1992). Each had its own performance characteristics but, crucially, software (tapes) produced by or for one standard could not be played on another. Moreover, as the installed base for a standard increased, the benefits to the users of machines employing that standard also increased (e.g. via a wider choice of pre-recorded films). In essence, once a buyer has chosen a machine, s/he is tied in to that format. This essentially means that any new-to-market products would only succeed if they employed the dominant standard (VHS), and in consequence the network effects would outweigh product performance improvement effects.

In general, however, we may observe that new-to-market products may be successful by offering improved performance at a price that is relatively not too high, by offering lesser performance at a much lower price, or by offering different performance at a higher or lower price. In each case one might expect demand for the original product either to be reduced or to stay the same (if the price does not change), but overall demand for the product class to increase.

The other type of new products that we have classified is product innovations that are new to the firm and, by implication, neither original nor new to the market. In this case we might think of the new variant as the same as another product already on the market. Of course there may be issues of reputation and branding that make the new variant different, but basically a new-to-the-firm product will be very similar to products that are already being sold. The success of the new variant will then be a matter of whether it is cheaper than other offerings on the market; and, if it does manage to sell, this will be at the expense of the existing variants that it emulates. Total sales of all variants may be increased if the new variant undercuts existing variants.

Thus both new-to-market and new-to-the-firm products with relevant pricing may well increase the total sales across all variants. However, there is still the possibility that the effect will be countered by expectations of new product launches. For durable products (not necessarily for non-durables), the expectation of future launches of higher quality, different, or cheaper products may lead buyers to postpone purchase until a later date, in order to benefit from expected future improvements.

5.5 **Conclusions**

In this chapter we have discussed the demand that a new product will face. In doing so we have differentiated between producer and consumer products; durable and non-durable products; original, new-to-market, and new-to-firm products; and products that are horizontally and vertically differentiated. By looking first at an original, durable, producer good supplied by a monopolist, we have drawn parallels with the literature on the diffusion of new technologies and have emphasized how learning, differences between buyers, stock effects, order effects, and other effects impact upon the demand for a new product. Karshenas and Stoneman (1993) attempt to measure the strength of these different effects for a particular example (numerically controlled machine tools) and find evidence to suggest that each has empirical support. A common thread is that, as price falls over time, demand will increase, although expectations of falling prices may lead to the postponement of sales. The findings about producer goods are also relevant in the consumer context; but here perhaps stock and order effects may not be so strong. However, many consumer goods may not be durable, and it was argued that, with non-durable goods, risk and uncertainty are less important.

In the context of consumer goods we also addressed the issue of new suppliers offering further products on the market, the demand for those products, and the impact on the demand for existing products and for the product class overall. We distinguished between new-to-market and new-to-firm products and looked at both horizontal and vertical product differentiation. To repeat here our earlier formulations, new-to-market products may be successful by offering improved performance at a price that is not too raised, by offering lesser performance at a much lower price, or by offering different performance at a higher or lower price. As a general result, one might expect demand for the original product to be reduced, or at least to stay the same but overall demand for the product class to increase. New-to-firm product innovations will be very similar to products already sold on the market, and sales of the new variant will then be a matter of price and will be at the expense of the existing variants. Total sales of all variants may increase if the new variant undercuts existing variants. Thus both new-to-market and to-firm products with relevant pricing may increase the total sales across all variants.

In general, the demand for a product innovation may change over time as products, knowledge, and the number of suppliers changes. One may expect prices (and price expectations) to play a major role in the determination of demand, but many other factors come into play too.

6 Capacity Creation, Pricing, and the Promotion of Product Innovations

6.1 Introduction

In this chapter we are mainly concerned with decisions relating to the pricing, promotion, creation and location of capacity to manufacture, goods that embody product innovations by those who supply such goods to the market. The discussion of decisions as to whether and when to generate such innovations is postponed until chapter 7. As in chapter 5, our emphasis here is upon products that are produced in significant numbers. We will not address products that are specially commissioned. The new products may be goods or services and may be in the creative sector or in other sectors of the economy. They may be producer goods or consumer goods.

Decisions on pricing, promotion, and capacity are interdependent, for example pricing low in order to generate high demand is only rational if there is capacity to supply the market. Thus, although we may, for expository reasons, discuss each topic separately to some degree, the interactions between them must not be forgotten. The other main point we must make is that issues of pricing, promotion, and capacity will be much affected by the presence or absence of competing suppliers to the market. In addressing the issues we thus consider not only the case where there is a monopoly supplier of an original product but also markets where other new products exist, both new to the market (but not original) and new to the firm (but not new to market). Most of our discussion is based on the premise that suppliers are profit seekers and perhaps profit maximizers. Their decisions are thus driven by the profit motive. They value profit tomorrow as well as profit today; however, tomorrow's profits are discounted at a rate that is determined by the firm itself. We will also assume that firms are risk-averse and, in order to adopt greater risk, must see the possibility of increased returns.

6.2 The monopoly supply of an original product

6.2.1 *Pricing*

The monopoly supplier of an original product may price his/her output without concerns about retaliation or competition from other suppliers. In the case where the product is non-durable (see Stoneman, 2011 for the context of soft innovations, but the arguments are the same for all products), we consider a simple model that assumes the following: (i) if a buyer purchases a product variant in a period, s/he will only buy one unit of that variant and no further units of it or other variants in that period (suitable examples might be ice cream, prepared foods, or a theatrical event); and (ii) there are M potential buyers or consumers on the market. If the monopolist supplies only one product to the market, then the probability that consumer j will demand this product will decline (i) as its price increases; (ii) as the inherent utility contribution of the product class declines; and (iii) as the marketed variant specification differs further from the ideal of the consumer. The overall demand for the product will then depend upon the number of potential purchasers, the distribution of consumer tastes and preferences, and prices.

A profit-maximizing supplier of this single product variant would equate marginal cost and marginal revenue. Thus, *ceteris paribus*, the lower the production cost per unit of output, the lower the variant price and the greater the sales will be. The lower the unit (marginal) costs, the lower the price will be, and hence the more distant the ideal specification of the marginal purchaser will be from the specification of the variant x upon the market. One might also argue that, if unit costs fell over time, then consumer acquisition of the product will increase, demand extending to consumers whom the product variant suits less and less.

As argued in chapter 5, for a durable good, if heterogeneous buyers have knowledge of the existence and characteristics of the new product, then demand at a point in time will be dependent upon the price charged at that point in time; and only if price falls over time will there be continuing demand for the new product (except for replacement demand), because all those for whom the product is desirable at the current price will buy today, leaving zero future demand at today's price. The supplier thus has the possibility of setting a low price today and exhausting the market today or pursuing a less extreme strategy of charging a price that starts high today but falls over time, total demand being spread over a period rather than compressed into a short period. Note that in the absence of competition the monopoly supplier may make this decision without fear that other suppliers would remove buyers from the market.

The monopolist may gain from postponing sales, insofar as (i) a falling price schedule enables some intertemporal price discrimination, as early adopters (who presumably can gain most from the new product) pay more than later adopters (who presumable can gain less); (ii) it is possible that the cost of production of the new product will decline over time (perhaps with some progress in process technology), which enables greater profit to be made, as sales are postponed rather than undertaken today and this reduces discounted total costs, although profits realized later are subject to a greater discount; (iii) spreading demand over time allows the market to be supplied with smaller production capacities operating for a longer period.

The supplier may not, however, have it all his/her own way. If it is known that the price of the new product is going to fall over time, buyers may decide to postpone their purchase until the price has actually fallen, which limits the extent to which the supplier can benefit from intertemporal price discrimination. Ireland and Stoneman (1986) discuss this matter and argue that, if buyers are myopic (do not foresee future price reductions), the supplier will maximize discounted profits by charging each buyer a price equal to the gain they will make from the new technology, the price falling over time and inducing new buyers until buyers can no longer be supplied at a price above cost. However, if buyers have foresight and in particular can predict the reductions in price over time, then they will wait, and as result the profit-maximizing monopolist will in each period charge a lower price partly counteracting the impact of increased waiting on sales, although the end point is the same.

If potential buyers do not have knowledge of the existence and characteristics of the new product, then the high-then-low pricing strategy may not be optimal. In particular, when information is self-propagated, the greater the number of early users, the greater the demand for the new product will be in a later period. In such circumstances (see Glaister, 1974) it may be optimal for the supplier to have lower prices in the early period in order to stimulate demand later, and then, at a later date, to have higher prices in order to increase profit. Although the discount factor may still have a role to play, clearly this is quite a contrast to the high-then-low price-discrimination strategy.

Of course, if the original product is not durable, then there is much less of a rationale for intertemporal price discrimination—postponing acquisition to a later date has no particular benefit. Lowering price at a point in time may generate more customers and more sales to each customer but will not of itself cause a customer to postpone purchase from today until tomorrow. In such circumstances the ability to spread knowledge via self-propagation may be the dominant factor affecting intertemporal pricing by the supplier, and so the profit-maximizing price may start low and then increase over time. But, again, a positive rate of discount may limit this strategy.

6.2.2 *Promotional activity*

The monopoly supplier of a new product may undertake promotional activity such as advertising, marketing, or product awareness activities, in the secure knowledge that, given the absence of competitors, that supplier will obtain the full reward of those activities, without spillover to other suppliers.

If promotion is to occur, the decisions of the suppliers will relate to how much promotion to undertake and when to undertake it (Glaister, 1974). The demand for some products may be more responsive to promotional advertising than the demand for others. One might for example expect sales to business to be less influenced by promotion or advertising than sales to consumers. Kaiser (2007) suggests, on the basis of an analysis of data on German service sector firms, that many product innovators do not need to advertise. In general, one would expect the amount of promotion to be greater when the elasticity of demand with respect to such expenditures is larger.

The issue of when the promotional activity should be undertaken is at least partly dependent upon how information spreads in the market. If information is self-propagating, then early advertising expenditure may encourage further information spreading, whereas with advertising later in the life of the product this may not arise. Later advertising may be of a persuasive rather than informative nature, attempting to influence buyers' preferences rather than their knowledge. The optimal strategy may be informative advertising early and persuasive advertising later. One might also expect that the nature of promotion will change over time. In the early stages of the life of a new product it may well be that existence and concept are emphasized. Later on performance and characteristics would come to the fore. The optimal intertemporal pattern of advertising for an original product by a monopolist may also depend upon the nature of the product. For products where there is a significant amount of self-propagated learning, there will be an incentive for the supplier to do the promotion early in those products' lives in order to stimulate such learning. For other products where there is less self-propagated learning, there is less incentive to front-load the promotional activity. Promotion might then be used to shift demand over time so as to increase that demand when costs are relatively low. This would suggest that promotional activity is preferable at the end of the product's life.

6.2.3 *Capacity creation*

If a firm is to supply the market with a new product, then it is necessary that capacity exists to enable production to occur. Although existing capacity may be transferred from producing an old product to the production of a new product, there is some evidence that, when firms introduce new products, they also introduce new production capacity. For example, in the UK

Innovation Survey of 2015, Table 3 (see DBEIS, 2016), it is estimated that the acquisition of capital represents 36.4 per cent of the cost of innovation (R&D representing less at 35 per cent).[1] The first main issue concerns how much capacity to create.

One may well expect that capacity is a durable good and as such will enable production over a number of years. However, demand for the product may not be constant over time (in fact demand may be related to the firm's intertemporal pricing and promotion strategies) and, as capacity is costly to create, firms do not wish to have capacity standing idle for periods of its working life or to be in a situation where it is unable to supply the market that its pricing and promotion strategies have created. There is evidence that the demand for new products (taking into account learning and pricing mechanisms) starts low and then grows at an increasing rate up to some point of inflection, after which ownership continues to grow, but at a decreasing rate (this is a basic finding of the literature on technological diffusion; see Comin and Mestieri, 2014). This suggests that, in terms of capacity, what is generally required for new products to match demand is an initially low or limited level of capacity that grows over time up to some point, after which no further capacity would be added and instead capacity would be reduced (perhaps through non-replacement of worn-out capacity), until the new product is fully diffused. Demand matching thus suggests that production capacity for new products would match the observed change in the level of product ownership (i.e. the flow of new owners).

However, there is an expectation that production processes may become more efficient over time, as capacity installed later is either cheaper to acquire or more effective in operation; hence the firm may wish to postpone investments in capacity, tending to encourage the spreading of capacity creation over time, or at least to scrap early installed capacity that is to be replaced by later capacity. In addition, learning by doing, in which the effectiveness of the production process increases in proportion to the amount that is produced, may also encourage firms to delay creation of new capacity until the best design and operation of such capacity has been tested and explored.

Overall, therefore, there are several factors that may influence the desired pattern of intertemporal capacity creation. Perhaps of the greatest relevance among them is the fact that decisions on capacity, pricing, and promotion are interdependent. Intertemporal pricing, which has a significant impact upon intertemporal demand, will depend upon production costs (among other factors). However, production costs will depend upon the time path of capacity creation, which will be dependent upon the time path of demand (among

[1] Surprisingly, Skuras et al. (2008) find nevertheless that, in a sample of SMEs in six European countries, the presence of product innovation reduces the propensity to invest in fixed assets.

other factors). Because the costs of creating capacity, the resulting costs of production, the prices charged, and the promotional expenditures incurred will jointly determine the profitability of supplying the new product over its lifetime, it is clear that decisions on capacity creation, pricing, and promotional expenditures need to be made simultaneously.

It is also obvious that many decisions regarding capacity creation will be based upon expectations of the market rather than the actuality of that market. Thus how firms form expectations upon the form of the demand curve for their new product, both in terms of position and slope, will translate back into decisions about what capacity to create and when. One might postulate that firms that are less optimistic in terms of future demand will create less capacity. The inability to be certain of the market is another factor that underlies the uncertainty embodied in capacity creation decisions. One might also postulate therefore that risk-averse firms will, *ceteris paribus*, be less likely to create large amounts of capacity.

6.2.4 *The location of production capacity*

A supplier of a new original product may supply the market in a number of ways. The classic analysis concerns investment in plant and machinery (or the transfer of plant and machinery from other uses) in the economy in which it is located. That capacity may also be used to supply export markets. Alternatively, the firm may invest in plant and machinery in order to supply the domestic market; but it may also, either wholly or partly, undertake overseas direct investment in order to create capacity to supply (perhaps individual) overseas markets. Further, a supplier may contract out production to a third party (in the domestic market or elsewhere) and effectively outsource its product. In this case the third party will have to create the production capacity.

A well-documented example of how manufacturers design new products in one country but produce those products overseas is that of Apple. Duhigg and Bradsher (2012) state that, although designed in the United States, almost all of the 70 million iPhones, 30 million iPads, and 59 million other products Apple sold in 2011 were manufactured outside the country. Although components differ across versions, all iPhones contain hundreds of parts, an estimated 90 per cent of which are manufactured outside the United States. Advanced semiconductors come from Germany and Taiwan, memory from Korea and Japan, display panels and circuitry from Korea and Taiwan, chipsets from Europe, and rare metals from Africa and Asia. And all of it is put together in China. Indicative of the extent of outsourcing in Apple is that, although Apple employs 43,000 people in the United States and 20,000 overseas, almost 700,000 people engineer, build, and assemble iPads, iPhones, and Apple's other products in Asia, Europe, and elsewhere—at factories also used by other electronics designers. In other words Apple not only has its new products manufactured overseas but has contracted out production to other companies. Such overseas production and outsourcing patterns, it is argued, has become common in hundreds of industries, including accounting, legal services, banking, auto manufacturing, and pharmaceuticals.

What factors, then, determine the location of production? There are many factors (for more detail, the reader is referred to an international trade textbook such as Feenstra, 2015). By comparison to the strategy of manufacturing in the domestic economy in order to supply home markets and of exporting to overseas markets, relevant issues include:

(i) Manufacture overseas may be cheaper for firms in the United States and other developed economies, especially as regards labour costs. US and western wages, for example, tend to exceed those of other potential manufacturing locations (for empirical evidence, Feenstra and Hanson, 1996).
(ii) Manufacture in local markets will minimize transport costs.
(iii) Home markets for labour may not be able to supply the number of skilled workers that are required.
(iv) There may be tariffs that make the price of home-produced goods exported to certain markets excessive and production in those markets may overcome such barriers.
(v) Production overseas can make the technology susceptible to appropriation by others or can stimulate local competition (or both), generating more challenging future markets.
(vi) The regulatory environment, tax regimes (especially as regards the bringing home of profits), and political stability are also factors that will be taken in to account in the choice of production locations.

The choice of whether to outsource production to another producer will also be influenced by a number of factors:

(i) If production is wholly or partially outsourced, the firm does not itself need to create as much new manufacturing capacity, either in its domestic economy or in overseas economies. Capacity creation and investment are, at least partly, undertaken by others. This reduces both the demand for capital resources and the risk attached to the new product.
(ii) The outsourced supplier may have manufacturing or production skills that are not available to the generator of the new product and, by outsourcing, the designing firm may overcome this problem.
(iii) The outsourced supplier may have knowledge that complements that of the lead firm and outsourcing enables the exploitation of such knowledge.
(iv) Sourcing supply from others may be cheaper than manufacturing in house.

(v) Certain countries, for example China, limit involvement in their markets to those firms that are willing to become involved in joint ventures with local firms—which encourages collaboration and outsourcing.

(vi) Outsourcing production to another company may lead to the loss of intellectual property, with consequent disadvantages.

6.3 **New-to-market product innovation**

In this section we explore issues relating to the pricing and promotion of product innovations and production capacity creation in markets where subsequent new product launches have created several different products and/or suppliers. Thus far we have considered demand for an original product with a monopoly supplier. Over time, however, other new to market products will be introduced either by the same or by other suppliers. Under monopoly, the impact of the new product launch upon the profits of the incumbent supplier are internalized by the monopolist. This contrasts with the situation where the new product variant is supplied by a competing firm: in that case some of the impacts fall upon rivals. These different distributions of impacts will significantly affect supplier behaviour.

We consider separately the cases where products are durable and non-durable and where the new-to-the-market products are horizontally or vertically differentiated (either better or poorer). Later on we consider the case where the products that are placed upon the market are new to the firm but not new to the market.

6.3.1 *The monopolist as new-to-market innovator, horizontal product innovation*

The base case is one where new products are put on the market so that there is more than just the original product available, but the extra products are generated by the same supplier as the original. This case means that there is no competition from other suppliers, and thus the innovator still has monopoly powers; but the monopolist is now a multi-product monopolist. The first decision that the monopolist has to make is how to price the different products on the market. Essentially there is no clear answer to this. The price will depend both upon the nature of the product and upon the nature of the innovation.

In a simple model of non-durable products, a single-product monopolist would equate marginal cost and marginal revenue. Thus, *ceteris paribus*, the lower the production cost per unit of output, the lower the variant price and the greater the sales will be. The lower the unit (marginal) costs, the lower the price will be, and therefore the more distant the ideal specification of the marginal purchaser will be from the specification of the variant already on the market. One might also argue that, if unit costs fell over time, then consumer acquisition of the product will increase, demand extending to consumers whom the product variant suits less and less.

If the monopolist launches onto the market a second product, which is horizontally differentiated from the first, then, at the same prices, if the product is going to be bought, the consumer will buy the variant that comes closer to meeting his or her own preferred specification (this is a property of horizontal differentiation). The changes in the pattern of demand that result from the launch of the new product arise from changes in the behaviour of two groups of consumers: those who would not have bought the original product but will now buy the new one; and those who would have bought the original product but now buy the new one instead. Other factors being equal, total sales, at a given price, are affected only by the numbers in the first category. Consumers in the first category are those for whom the characteristics of the original product were too distant from their preferences. The better match between product characteristics and consumer preferences allows the monopolist either to sell more at a given price or to charge a higher price for a given number of units sold.

One might thus expect that, under horizontal product innovation by a monopolist, if consumer preferences are uniformly distributed and the costs of producing the new and the old variant are the same, then the prices of the new and the old variants will be the same. If the pricing rule is the equating of marginal revenue and marginal costs of production, then, if unit costs of production do not change as a result of launching a new product, the impact of such innovation upon the common price of the two variants will depend on the impact upon marginal revenue. As consumer preferences are more adequately met when there is more than one variant on the market, one may expect that marginal revenue will increase for any given number of units sold and that, in consequence, prices will be higher the more variants the monopolist places upon the market.

An alternative case is one where the product is durable rather than non-durable, that is, where the consumer enters the market in each period until a unit is purchased, after which the consumer buys no further units. In such a case the single-product monopolist will price the product over time so that in each time period the price will be that which, for the marginal consumer, equates the benefit of waiting one further period before buying the new product equal to the marginal cost of producing the product today rather than tomorrow. Models of this kind predict that (for a given number of consumers) the price of the product will decline over time (until production costs stop falling: at that point the market closes), consumers obtaining less and less benefit as time goes on. If, however, the monopolist differentiates the product horizontally, by making an additional new-to-market product variant available, then one might expect that buyers would be willing to pay a higher price or buy at an earlier date, since for some of them a product that suits their preferences better will become available. One may thus also expect the price charged by the monopolist to be higher at all points in time.

Another possible scenario is one where the product is durable but may be a repeat purchase, although each purchase may be of a different variant (examples are books, DVDs, or CDs). In such cases the consumer will probably buy many variants but, as the product is storable, will not purchase a variant a second time, and any later purchases will be of different variants. For such products, the consumer enters the market in every period but will never repurchase the same variant previously purchased. Stoneman (2011) argues that, in such a market, if there is only one supplier and only one product variant (the variant has an infinite life), its price will fall over time, as the costs of supplying the variant fall over time. As prices fall, sales extend to consumers for whom the variant is less preferred. As the costs of production fall, further sales will take place at lower and lower prices, the product appealing to buyers who have less and less preference for the variant, until costs fail to fall any further; after this the product will no longer be available. If the monopolist launches a new variant on the market, the pricing of the new (and of the existing) variant will be dependent upon the cross-price elasticity of demand. However, as variants age, their potential sales will decline. This suggests that, as time proceeds, any impacts upon the sales of new products arising from existing variants would be getting smaller. For a monopolist, this suggests that pricing of the new will be less constrained by the existing products as time proceeds.

6.3.2 *The monopolist as new-to-market innovator, vertical product innovation*

Vertical product differentiation concerns products that come at different agreed quality levels. A useful example is rail travel, where there may be first and standard class, with an agreed quality ordering (and, of course, with different prices). Classic literature references are to Shaked and Sutton (1982, 1983). Vertical product innovation will involve the introduction of a new product that is agreed by the market to be of different quality from existing products. The quality may be higher or lower.

For the monopolist who launches a new, vertically differentiated variant, one will expect that the new and the old variants will have different prices, that of the higher quality variant being above that of the lower quality variant. An interesting issue is what happens to the price of the existing variant when a monopolist launches a new, vertically differentiated variant on the market. One might argue that, if the new variant is of higher quality, its price will be higher and it will reduce the sales at the current price of the existing variant, whereas if the new variant is of lower quality its price will be lower and, again, will reduce sales of the existing variant at its prelaunch price. However, the exact pattern of prices will depend upon both fixed and variable costs, as well as upon the distribution of buyer preferences and incomes. Sutton (1986) argues that, if all consumers have identical tastes but different incomes and consume only one unit of a product, and if the costs of production increase sufficiently quickly with product quality, then product prices will also increase with product quality. Then consumers will partition themselves by income in such a way that brands of successively higher quality are purchased by consumers in successively higher income bands, which reflects the fact that a consumer's willingness to pay for quality improvements is an increasing function of income.

6.3.3 *The monopolist as new-to-market innovator, product promotion and capacity creation*

Promotional activity of a new-to-market product by the monopolist supplier of an original product innovation will be governed by two factors. First, promotion of the new-to-market product may increase demand for that product but this may be at least partly at the expense of the monopolist's existing products. Such promotional activity may thus have some downside. However, it may also stimulate demand for the product class in total (both the original and the new-to-market product)—a benefit that the monopolist will be able to internalize. This contrasts with the case where the new-to-market product comes from a different supplier. One may expect that, with the launch of a new-to-market product, if only for informational purposes, the monopolist would undertake greater promotional activity, although this could be limited as a result of the impact upon the original product offering. We have discussed, in the case of an original product, how promotion may be spread over time depending on the learning processes involved in the market. We see no reason to change those arguments in the case where the monopolist has more than one product offering.

We have argued that, basically, if a monopolist launches an additional new-to-market product, then joint demand for the product class will be greater than before, either because the new-to-market product will more satisfactorily meet the needs of buyers with preferences more distant from the characteristics of the original product offering or because it will enable coverage of a wider spread of the income spectrum. This increase in demand (either ameliorated or encouraged, depending on the circumstances, by price changes) will encourage the creation of extra production capacity. One might thus expect that the launch of a new-to-market product by the monopolist will lead to further investment in capacity (either in house or from other suppliers).

6.3.4 *The monopolist as new-to-firm innovator*

One might think that the original supplier of a new product would have no incentive to launch a new product variant that is the same as the existing product variant—and hence the monopolist would not undertake new-to-firm (but not new-to-market) innovation. However, if a new product that is identical with the firm's existing product can be produced more cheaply, then the monopolist may have a profit incentive to launch that product, remove the existing product from the market, charge less, and sell more.

6.3.5 *Many new-to-market innovators: horizontal product differentiation*

The monopolist model has the advantage of introducing approaches and concepts while also being informative as to outcomes and causes. It is, however, necessary to get away from the monopoly assumption and to consider more appropriate scenarios. In particular, we are interested in situations where, although there is a single supplier of the original product innovation, over time other firms enter the market and offer their own variants on the original product. These variants may be differentiated vertically or horizontally or be new to the firm and not to the market, and thus the same as other product offerings. We will talk of each possibility separately. Once again, however, it matters whether the product being offered is durable or non-durable.

An influential contribution to the literature in this field is by Dixit and Stiglitz (1977). These authors consider a world where products are non-durable and each product offering is horizontally differentiated from others. Consumers are allowed to buy more than one unit and consumer preferences are equally distributed across product space. There are many different suppliers, but each supplier is assumed to produce only one variant, the variants being equally distributed across product space. Once there is more than one product available, the market scenario is one of monopolistic competition.

Assuming that each firm aims to maximize profits as all other prices are held constant, for each firm, at the profit maximum, marginal revenue will equal marginal costs. Assuming that each firm has the same common marginal costs of production, all firms will charge the same price. Dixit and Stiglitz (1977) show that this price will be higher the greater the marginal costs and the lower the price elasticity of demand. The important point, however, is that, as the number of firms increases, the price in the market will be lower. This may be taken as indicating that, as the number of new-to-market entrants with horizontally differentiated product offerings increases, the price in the market will be lower and both demand and quantity supplied will be greater. The lower price will also mean (given common marginal costs) that the profits of the suppliers will be lower the greater the number of new-to-market suppliers. Assuming that positive profits are an attractant to further new-to-market product launches, one may also argue that, with free entry, at the limit, the number of new-to-market launches will be such that there will be zero profits (the number of firms and variants in the market being a function of the elasticity of demand, the marginal costs of production, and the level of fixed costs), with a price such that total revenue will equal the sum of production costs and the fixed costs of the product launch.

The durable product case will exhibit similar characteristics as the non-durable product case, although there is a particular overtone that may be of some importance. This is most clearly highlighted in the case where the durable product is to be purchased only once. Then a sale by one firm in any time period reduces the overall market for all other firms by one unit in any future time period. Thus, when there is more than one supplier in a market such as this, the impact of product proliferation will be, at least partly, externalized. This leads to what is known as 'common pool effects'. When there are common pool effects, suppliers may prefer to make a sale today at low profit margins rather than allow others to make that sale and to lose the sale for all time. In such markets it is commonly argued that increasing the number of suppliers leads to lower prices and encourages earlier purchase of the product. One may thus confidently argue that an increasing number of suppliers, each with its own single-purchase product, will cause each to charge a lower price. Thus new-to-market innovation by suppliers other than the original product suppliers in markets with durable products will lead to lower prices and will also bring demand forward in time by comparison to a world with a single supplier.

6.3.6 *Many new-to-market innovators: vertical product differentiation*

Sutton (1986) argues that the basic determinants of the number of vertically differentiated products that may be launched on a market are the nature of technology and tastes—specifically, the relationship between consumers' willingness to pay for quality improvements and the increase in unit variable cost associated with such improvements. In the absence of fixed costs of product diversity, a number of new-to-market products may be launched by different suppliers such that there will be a number of products upon the market each serving a different aspect of the income spectrum. Fixed costs relating to product variation will determine the extent to which groups of consumers are serviced by a product—as opposed to individual consumers. It is reasonable to argue that, as the number of new-to-market suppliers increases, consumers' demands will be more equally matched and total demand will be greater. At the limit, however, the number of varieties launched will be such that the last product launched is just profitable and a further launch would not be profitable. That is to say, as vertical product innovation is extended via new-to-market product launches, the profits of all suppliers (including the originator) will fall.

6.3.7 *Many new-to-market innovators: product promotion and capacity creation*

When there is more than one supplier of a new product, promotional activity will have two impacts. The first is to increase the market share of the promoter's own product offering, which occurs partly at the cost of the other suppliers in the market. The greater the promoter's original market share, the less it is externalized. Secondly, promotional activity will influence the demand for the product class, some of the benefit of which will be reaped by competitors in the market rather than being internalized. Whether product proliferation via new-to-market products encourages more promotional spending or not depends upon how the balance of these effects works out as the number of product suppliers increases. Compared to the monopoly case, the market share effect will encourage more advertising in oligopolistic markets, whereas the product class effect will encourage less.

As far as capacity is concerned, one might expect that, as there is more new-to-market product innovation, the demand for the product will increase and hence the amount sold will also increase. This will mean that more innovation will encourage capacity creation. However, to the extent that innovation implies that existing firms lose market to new entrants, expectation of this effect may mean that existing suppliers are more averse to creating new capacity.

6.3.8 *New-to-firm (but not new-to-market) innovation*

Having considered original innovation and new-to-market innovation, we may say a few words about new-to-firm innovation. The essence of this kind of innovation is that a new product is placed on the market that is essentially the same as a product that already exists. It is possible that the newest market offering is cheaper to produce than the existing product (which may be the case if it is introduced by the current supplier of an identical product). The new product, being new only to the firm, is neither vertically nor horizontally differentiated from at least one product that is currently available. One may expect that such new-to-the-firm innovation will lead to a lower market price for the affected part of the product spectrum, either because of competition or because of the lower costs of the new product. It is possible that, if the entry is based on lower costs, the existing variant will be driven out of the market. We may thus associate this kind of new-to-the-firm innovation with subsequent lower prices, higher demand, and perhaps some turnover in the identity of products and product suppliers.

6.4 Conclusions

In this chapter we have addressed a number of issues related essentially to the decisions of the suppliers of new products on pricing, promotion, capacity creation, and the location of production, primarily assuming that firms maximize profit. We have looked at variants related to durable and non-durable products, monopoly and oligopoly, horizontal and vertical differentiation, and original, new-to-market, and new-to-firm product innovation.

The thrust of our findings is that firms may well have incentives to further innovate after the launch of the original product, although there are limits to the process—depending on fixed costs, variable costs, and the elasticity of demand. Further innovation, either by the original suppliers or by others, tends to lead to lower prices and greater demand and output. This may also involve greater capacity creation (in some locations). We find, however, that the incentives, for the originator and for potential new entrants, may differ according to the internalization and externalization of the impacts of their behaviour. Prices will tend to be lower when there are many suppliers than when there are fewer. This means that the quantity supplied will be higher if there are more innovators. Capacity creation may also be greater. We have not committed ourselves on the matter of promotional expenditures.

7 The Incentives to, and Constraints upon, Product Innovation

7.1 Introduction

The main aim in this chapter is to explore the factors that affect the firm's decision to undertake product innovation. There is an underlying presumption that firms or entrepreneurs (which for present purposes are largely considered as synonymous) are profit seeking (if not profit maximizing). The profit from innovation will be determined by the (discounted) net revenues earned from the supply of a new product and the costs of developing and placing that new product on the market. As we shall argue, there are many ways in which product innovation and profitability interact. There are also a number of triggers that might initiate product innovation in the search for profits, and thus to talk of profit seeking is only part of the story. The search for profit is also taking place in an uncertain world; thus there is a risk element to innovation that has to be taken in to account in the innovation decision.

Our discussion will take place in the context of the issues addressed in chapters 5 and 6, which relate to the demand for product innovation and the pricing and promotion of and capacity creation for the supply of innovative products. Basically, in addition to the costs of developing product innovations, these are the key elements impacting upon the profitability of product innovation. As an indication of the different costs involved in getting new products to market, data from the 2015 UK Innovation Survey (see Department for Business Innovation and Skills [DBIS], 2016b) indicate that, as a percentage of the total spend on innovation (unfortunately for our purposes, this total encompasses both product and process), the acquisition of advanced machinery equipment and software accounted for 36 per cent, internal R&D accounted for 35 per cent (with external R&D another 4 per cent), all forms of design accounted for 9 per cent, and the market introduction of innovations accounted for just over 10 per cent, the balance (6 per cent) being expended on training and the acquisition of external knowledge.

In the following sections we first address the general question of why firms or entrepreneurs might undertake product innovation. From there we move to consider the extent of the innovation (a concept yet to be defined) that firms might undertake before analysing the type or direction of innovation that can be expected and the timing of innovation activity. Finally we address issues pertaining to the risks of product innovations and to the potential constraints to innovative activity, considering, inter alia, the financing of product innovation.

7.2 **Why undertake product innovation?**

There are several classifications of the reasons why firms innovate. For example, the reasons are often considered to be related either to technology push or to demand pull. We do not explicitly pursue this classification here but offer some arguments that might be classified as demand push and technology pull. Alternatively, there is a classification that we have found useful in the past; it was proposed by Freeman (2000), who argued that innovation may be offensive (attacking the market share of others), defensive (a reaction to loss of market share to others), or protective (an insurance against innovation by others). At other times we have found this classification useful, but in the present context it is rather constraining and thus we have decided against using it. Instead we offer a list of factors or situations that may stimulate product innovation without trying to provide a higher level of classification. Our approach may be seen as based upon the view that product innovation is a search for profits that is subject to various starting triggers; and what we try to do here is list a number of the potential triggers.

Perhaps the classic idea about what drives product innovation is that a firm observes a market opportunity and, on the basis of either an existing or a newly developed technology, supplies a new product to the market in order to capitalize upon that opportunity. The opportunity may be triggered in such a way that the new product is original. In that case the costs of developing the new product are likely to be greatest, insofar as the development process is likely to involve activities not previously undertaken. Alternatively, the market opportunity my be triggered by the new product launches of other firms, in which case the new product (offensive innovation) is unlikely to be original but can be either new to the market or new to the firm. The former is likely to involve greater expenditure upon new product development than the latter. One could label these factors 'market pull'. Of course the new products will require perhaps costly development, and hence such market opportunities will not always be profitable.

This may not, however, be the only route by which the market leads innovation. One alternative is when the market for a firm's existing product has been exhausted. This could be common in cases where the product is durable and is only purchased once by any consumer. Once all consumers have the product, there is no new demand and the market is exhausted. The firm needs to launch a replacement product to continue its activities. Very good examples exist in the creative industries. Firms may need to launch new music recordings, films, books, and video games because the number of new purchasers of existing offerings has already dwindled or is expected to dwindle, and only new products will maintain sales. A variation on this theme is that, as the number of potential buyers declines with increased prior ownership, the profitability attached to the price at which existing products can be sold declines and profit margins decline too. This makes the potential revenue from a new product relatively more attractive. One could label this 'market push'.

A firm already in the market may experience loss of market not only because the market for its existing product has been exhausted. Loss of market could come through the launch of a superior, different, or cheaper competing product by another firm. The loss of market may then cause the firm under consideration to react by innovating in turn (defensive innovation). The prospect of such an occurrence may also lead a firm to undertake (protective) innovation, either in order to react quickly or to head off such innovation by others. An incumbent could even acquire a challenger (via merger or acquisition) rather than openly competing with this entrant, and might even sell simultaneously both the existing products and the new one (see Gabszewicz and Tarola, 2012).

In addition to these market factors, triggers for new product launches can come from the technology side. Classically, technological advances can make original new products either feasible or economically viable. Thus for example advances in electronics have enabled the development and marketing of many new IT-related products (mobile phones, laptops, etc.) at prices that are low enough and with performance that is good enough to encourage consumer purchases. Over time, further technological advances have improved the performance of such products, changed their functionality, and reduced their cost. One may note that advances in process technology may be a trigger to product innovation in addition to advances in product technology. Advances in process technology may enable the supply of cheaper products, but those products may have to replace existing products (an example is the replacement of iron and steel by plastics-based materials in various products).

Such arguments may seem to suggest that much product innovation is led by changes in technology. However, to some degree at least, many of the technological advances are endogenous. Potential market opportunities may lead firms to expend resources on developing new technologies that may be employed in meeting those opportunities. Although there may be advances in knowledge that are not driven by (at least specific) market opportunities, often labelled basic research, such research is largely undertaken outside the commercial sphere, in universities or other public research institutions, is financed by government, and represents only a small part of the total activities in science and technology. Most expenditure on science and technology can be labelled applied research or development spending, having a specific objective in view. This would suggest that most technological development activity is endogenous and thus new technological opportunities are not often the basic driving force behind product innovation.

We prefer the view that there is little mileage in disputing whether product innovation is led by market opportunities or is technologically driven. Basically the technological spending and development will be undertaken with an eye on market opportunities, while the market opportunities pursued via product innovation are likely to be chosen by the cost of the technological development required to meet those opportunities.

There are certain product innovations, however, that will not require expenditure on technological development but may be based rather upon personal creative activities. Some creative activities may be part of resource-intensive processes, for instance architecture, but others may be far less resource-intensive, for example the writing of a book. When the development process is less resource-intensive (and also when the costs of supplying the innovative product are low), product innovation may be less related to profit opportunities and more related to aesthetic preferences. When the creative process is also resource-intensive, although not necessarily technological (e.g. the restaging of an opera, or the recording of a new CD), one could well argue that the relationship between the market and development costs is still crucial in bringing the new product to market; however, the costs incurred are of a different kind. In cases where the costs of creation are high in relation to market demands, aesthetic arguments may be used to support public subsidies to the innovation activity (e.g. public subsidies for opera and national orchestras).

7.3 How much innovation?

We have discussed the generalities of the innovation decision, basically addressing what may trigger a firm to pursue product innovation. In this section we address the extent of product innovation that a firm may undertake. For these purposes we need some sort of measure of the concept of the extent of innovation. A number of possibilities spring to mind. When innovation is discussed in the media or by politicians, it is often R&D spending that is at the centre of attention. However, R&D has a number of problems as a measure of product innovation activity: first, it encompasses both product and process innovation and we are here really only interested in the former; secondly, it is a measure of the inputs to technology development and not a measure of the output of that process, whereas the output is what we wish to measure. There is some validity in measuring innovation by the resources expended on the activity and thus, to some degree, we follow the standard approach but, because of the first problem, we will talk of resources expended on product development rather than of R&D *in toto*. In doing this we consider that the greater the resources expended on product development the more product innovation a firm will undertake, be it in terms of the number of new products introduced or in terms of the extent to which those products are an improvement on the existing products on the market. However, there is also mileage in being more specific as to what the extent of product innovation means. Thus we also look at explicit models of horizontal product innovation, where we consider both the market 'niches' for which an innovator might aim and the number of new products that s/he may decide to introduce. Finally we examine vertical product innovation and consider both the number of new products a firm may introduce and the extent of improvement for which the firm might aim.

7.3.1 *A general approach*

Starting with a general approach, we consider that a firm needs to expend resources developing new products, and the more resources it expends the greater the extent of product innovation that it undertakes will be. In a standard economic modelling approach, however, we argue that there is declining marginal productivity in new product development in that, although extra resources mean more innovation, the increase in innovation generated by an extra unit of expenditure declines with total expenditure.

The extra revenue that is generated from product innovation will be dependent upon many factors, as detailed in chapters 5 and 6, for example the size of the market, the number of competitors, the expenditure on promotion and installing production capacity, pricing (these latter three are probably endogenous), and so on. If the net extra revenue to be earned from innovation is greater than the resources required for development, then the firm may be expected to undertake such innovation. Putting aside risk, a profit-maximizing firm will expend on product development up to the point where the expected (discounted) marginal extra revenue from such spending equals the expected marginal development cost. It is important to point out, however, that net extra revenue should be considered to be relative to a counterfactual. The alternative to innovation is not always the status quo. If others are innovating and or the world is changing, then not innovating may mean, for example, a fall in revenue. The nature of the counterfactual may well be very dependent on the competitive environment.

From the simple equality of marginal revenue and marginal costs one may argue that factors that affect marginal cost and marginal revenue will impact upon the development undertaken and the innovation introduced. Thus, for example, if the market is large, one might (other factors being equal) expect more innovation than if the market is small. If the costs of developing a unit of innovation is large, then one might also expect there to be less rather than more innovation. In addition, if the cost of producing innovations increases sharply with the number of innovations produced, then fewer should be expected rather than more.

One of the more enduring questions in economics, however, concerns how innovation is related to market structure. In essence this addresses whether larger firms in more concentrated markets will be more innovative than smaller firms in less concentrated or more competitive markets. A final answer to this question is still open. An interesting approach to this issue can be found in Lin and Saggi (2002), where product and process innovation are compared. In fact the question is deceptive, for it encompasses a number of questions in one. As for the impact of firm size, if the issue that larger firms means more concentration for a given firm size is put on one side, then the impact of firm size is largely a matter of access to resources. Large firms may well have easier access to finance (say, through higher cash flow, reserves, or easier borrowing) or to skills (through a bigger labour force), especially if they are part of a group, and this can make them more innovative, *ceteris paribus*. Larger firms may also be able to carry a loss more easily, and this can encourage innovativeness. As for the effect of concentration, putting the issue of firm size on one side, there are three main questions we isolate here. (i) If the new product is to be bought by other producers and thus represents a process innovation for them, will a concentrated or a non-concentrated buyers market offer greater incentives to the product innovator? (ii) If the new product is a consumer product, will an increase in the number of sellers increase or decrease the incentive to innovate? And (iii) will a greater number of potential innovators increase or decrease the incentive to innovate?

Implicitly assuming that innovators have enforceable intellectual property rights on their innovations, taking the first of these questions, the classic argument presented by Arrow (1962) is that, because the (monopoly) supplier of a new technology is able to price that innovation in such a way as to extract a premium equal to a monopoly rent in competitive markets but not in monopoly markets, competitive buyer markets encourage greater innovation. Taking the second question, as we have argued in chapters 5 and 6, as the number of competing suppliers of a new product increases, the price of that product will be lower, investment in promotion may be greater, and more investment in capacity could be necessary to meet early demand. This all suggests that more suppliers mean lower incentives to undertake product innovation, because there is a lower potential return. However, more suppliers spending fewer resources each might still mean greater industry spending on product development when summed over all suppliers. Finally, if there are more potential innovators, the counterfactual may involve more innovation by others and lower profits in the absence of innovation, thus encouraging more innovation. Alternatively, with more potential innovators, it may be necessary to get to the market before others. This common pool effect can mean that firms need to spend more resources on innovation, so firm and industry spending will be greater and product innovation will occur earlier.

7.3.2 *Horizontal product innovation*

The firm placing an original product on a market may locate that product in a product space where the potential gains from innovation are greatest. If consumer preferences are equally distributed in product space, this may be anywhere in that space. If preferences are unevenly distributed, then the product is more likely to be located where the concentration of preferences is greatest. The original product may not necessarily closely match the preferences of all consumers and further, horizontally differentiated products may be placed on the market—these are new to market but not original. For a monopolist supplier, if development costs are independent of product characteristics, the new products will be placed in product space in order to maximize revenue. Revenue will increase with the number of products on the market, primarily because, as more products are offered, consumer preferences may be more closely met and the price charged for both new and existing products can be increased. Although a monopoly supplier may suffer reductions in the quantity of sales of one product if the characteristics of the new products significantly overlap with those of the original product, the higher prices that can be charged may to some extent compensate for the loss and allow revenue to increase. One could also argue that, if buyer information on products is self-propagating, then more products with more sales will stimulate overall demand. Extra new products may be introduced until the extra net total market revenue generated from a new product launch is equal to the costs of launching that new product. That will determine the extent of innovation.

In the case where more than one firm is able to offer a product to the market, the picture will be different. The crucial difference arises from the fact that, if another firm offers a new product to the market, that new product may well take sales revenue from the original product. The balance of costs and revenues arising to the innovator from a new product launch will thus be different from in the monopoly supplier case. To illustrate the possibilities, we consider two classic approaches in the literature. The first is attributed to Hotelling (1929) and can be thought of as represented by ice cream sellers on a beach. The consumers are equally distributed along the beach. Each buyer will purchase one unit of the product if the utility gain from purchase exceeds its price plus travel costs. The utility gains from an ice cream are considered to be the same for all buyers, but travel costs differ across purchasers depending on where the product is located in preference space, that is, depending on its location on the beach. The travel cost will be low for customers located on the beach near the seller and high for distant customers. Thus, if two sellers charge the same prices, buyers will go to the seller nearest to their location.

A single seller with a single location will be located at the centre of the beach and charge a price that, for simplicity, we will assume induces all buyers to purchase an ice cream, the highest travel costs, incurred by those farthest from the seller, being not too onerous. This also means, by assumption, that those nearest the seller, who incur minimum travel costs, would be willing to pay a higher price. Let the monopolist have revenue R, production costs C, and fixed costs F. The monopolist thus makes profits $\Pi = R - C - F$. Assuming that the beach is of length 1, the buyers will travel on average a distance of 0.25 and, assuming a travel cost of 1 per unit of distance, we may specify the total travel costs to be 0.25N, where N is the total number of buyers. The monopoly supplier may, however, decide to have two outlets. These will be located so as to minimize the average travel costs of buyers, enabling the seller to charge the highest price and still sell ice creams to all. The locations of stalls that will achieve this are at one quarter and three quarters of the length of the beach from either end. Average travel costs will be half of the single location costs (0.125), and prices may rise by this amount. The extra revenue earned by the seller would then be equal to 0.125N. Assuming that operating costs double with two kiosks and that supplementary fixed costs, F, must also be incurred (the originally fixed costs being sunk), the supplier's extra profits will be $0.125N - C - F$. The single seller will thus find the opening of a second kiosk profitable if $0.125N > C + F$. A monopolist may continue to locate further outlets equidistantly on the beach until the gain in profits from so doing equals the fixed cost of establishing a further outlet. The buyers' travel costs are being reduced as the number of outlets increases, but buyers have to pay a higher price for ice cream to compensate.

Suppose, however, that the original seller is located centrally and another seller is considering opening business on the beach. Where will that seller locate him-/herself? We assume that both sellers charge the same price, so the consumers' choice of seller will be determined purely by the distance they have to travel. In this case it is argued that the new entrant will establish a kiosk right next to the original seller. In doing so one seller will supply the left-hand side of the beach, whereas the other will supply the right-hand side. The buyers will still be at the same distance from a kiosk. Why do the sellers not locate themselves as in the single-firm case, at one quarter and three quarters of the length of the beach from either end? Essentially, if they did so, then, by moving closer to the centre, each one could increase his or her market share at the expense of the rival, and such an arrangement is unstable (without collaboration). The original seller will lose half his market, with associated revenues and profits reduced. Assuming that the operating costs of each seller are C, the original seller will have profits (fixed costs being sunk) of (0.5R – C), the entrant, who has to bear the fixed costs F, having profits of 0.5R – C – F. Thus the original supplier will find that profits are reduced by entry. The new entrant has an incentive to enter the market equal to half the original firm's profits, minus fixed costs. The buyers make no gain.

A monopoly supplier facing a possible new entrant may (i) do nothing and allow the entrant to establish a position on the market, and the profits of the former monopolist would diminish from (R – C – F) to (0.5R – C – F), in other words they fall by 0.5R; or (ii) establish a second outlet and increase profits by 0.125N – C – F, which is a superior strategy if 0.125N – C – F > – 0.5R, in other words depends on operating and fixed costs. If costs are not too large, the latter strategy would be preferred. Essentially, even if the possibility of entry does not lead to new entry, it may prompt a monopolist to innovate. However, one might note that, if the innovator of an original product expects new-to-market innovation after his or her entry, then the expectation of reduced profits may deter the original innovation.

The argument just detailed, whereby a new entrant takes a place close in product space to an existing seller, has been used to suggest that, if new firms enter a market, then they will offer products with characteristics similar to those of products already on the market. For example, one observes that supermarkets are often located close to each other or that all designs of new-to-market products are often similar.

There is an alternative approach that we may use to explore this further and that, as a by-product, also copes more easily with more than one potential entrant. This alternative story is attributed to Salop (1979). The main difference from the first is that, instead of buyer preferences being assumed to be evenly distributed along a straight line, they are assumed to be distributed evenly around a (unit) circle. An example is a situation where, instead of a being evenly distributed on a beach, the ice cream buyers are equally distributed on the banks of a lake around which they must walk to reach the ice cream seller. Again, however, for a given price buyers try to minimize travel costs and will buy if travel costs (divergence from the ideal specification) plus the cost of the ice cream are less than the utility of having an ice cream. An initial original innovator may locate him-/herself anywhere on the circle. We will assume that buyers like ice cream enough to buy from the initial seller, the average travel distance being a quarter of the circumference of the circle (buyers can approach from either side). Let R be the revenue, C the operating costs, and F the fixed costs; the monopolist will make profits $\Pi = R - C - F$. Assuming that the circumference of the lake is of length 1, the buyers travel on average a distance of 0.25 and the total travel costs across all consumers will be 0.25N.

A monopolist supplier may open a second kiosk. If s/he does so, s/he will locate the kiosk diametrically opposite the first. The buyers in each semicircle will travel to their nearest kiosk, thereby halving the average travel distance (to one eighth of the circumference of the lake). The average travel cost will be 0.125 and the total travel costs will be 0.125N. By internalizing the reductions in travel costs through an increase in prices, the monopolist would increase profits to R + 0.125N – 2C – 2F from R – C – F, that is, an increase would be 0.125N – C – F. The monopolist will add a further kiosk and will continue to do so as long as the gain is positive. These kiosks will be equidistantly placed around the circumference.

Instead of the monopolist opening a new kiosk, a new entrant could make a new-to-market product launch. Unlike in the case of the beach, where this new kiosk was placed next to that of the monopolist, in this case the kiosk will placed diametrically opposite that of the originator (as if a monopolist were opening the new kiosk). This is because, if the new kiosk were placed next to the originator, then each buyer would serve half the market, the average travel distance being a quarter of the circumference. With diametrically opposed locations, average travel distance would be one eighth of the circumference, and thus higher prices may be charged. Other new entrants may also place new kiosks around the lake, all equidistant from each other.

As new entrants establish themselves, part of their return comes from reductions in the sales and profits of the existing suppliers—a reduction that does not occur for a multi-product monopolist. One might thus argue that there is greater incentive for new entrants to innovate than for the monopolist to do so. Of course, once one takes into account that expectations of further entry may impact on the probability of entry, this argument may no longer apply. One may note, however, that in this case sellers do not bunch in product space but instead are evenly located in that space.

These two simple stories suggest that, although product variants may group in preference space, this is not a necessary outcome. One might argue further that, if preferences, instead of being evenly spread across the population, are in fact more unevenly distributed, the incentives will be to encourage innovations servicing the thicker parts of the market, a smaller number servicing the thinner (e.g. pop music recordings vs classical music recordings). Further, if the production costs or fixed costs vary with the aspects of the product space that is serviced, or if there are scale economies (or diseconomies) of production, one might expect these also to affect the characteristics of innovations launched on to the market.

These stories are informative but are not particularly well suited to modelling the extent of innovation, that is, the number of innovations placed on the market. Another approach to the issues at hand that has been particularly influential in the economics literature was introduced by Dixit and Stiglitz (1977). The product is considered non-storable, consumers are allowed to buy more than one unit, there are many different suppliers, and the market scenario is one of monopolistic competition. Each supplier produces only one variant, and variants are equally distributed across product spaces, as are consumer preferences. All firms are considered to be the same (the market is symmetrical), each firm has common marginal costs of production, and all firms will charge the same price. Each firm is assumed to maximize profits expecting prices of all other firms to be held constant, and at the profit maximum, for each firm, marginal revenue will equal marginal costs. The industry price will be higher the greater the marginal costs and the lower the own-price elasticity of demand. If one assumes free entry, then the number of firms in the market—that is, the number of product innovations undertaken—will be determined, such that there will be zero profits (the number of firms and variants in the market is a function of the elasticity of demand, the marginal costs of production, and the level of fixed costs), with a price such that total revenue equals the sum of production costs and the fixed costs of product launch. Thus the extent of innovation is to be determined by the elasticity of demand, the marginal costs of production, the level of fixed costs, and freedom of entry.

One particular type of product is the single-purchase durable good. The key aspect of such products is that the consumer is assumed to enter the market in each period until a unit is purchased, after which the consumer buys no further units. A monopolist may have an incentive to introduce further product variants because, for example, buyers would be willing to pay a higher price or buy at an earlier date if a product more suited to their preferences is introduced, for which the price charged by the monopolist could be greater at all points in time. In fact the monopolist would launch new product variants if that introduction (relative to development and other such launch costs) (i) sufficiently brings forward the buyers' desire to purchase to increase discounted revenues; (ii) sufficiently changes specification to increase the total number of potential buyers; or (iii) sufficiently decreases the production costs of the product.

When there is more than one supplier in such a market, the fact that the impact of product proliferation is at least partly externalized may well generate common pool effects. Essentially, if the product considered is to be purchased only once, then a sale by one firm reduces the overall market for all other firms by one unit. In such markets it is commonly argued that increasing the number of suppliers leads to lower prices and encourages earlier purchase of the product; however, it is perfectly reasonable to also argue that the same effect may induce firms to undertake more horizontal product innovation, that is, a firm will introduce new variants to capture buyers before its (actual or potential) rivals can do so.

Earlier on we have also discussed products that are durable, but different variants are purchased over time (repeat purchases). Examples are books,

DVDs, or CDs. In fact a detailed analysis of product launches in the video retail industry can be found in Seim (2006). In such cases the consumer will probably buy many titles but, as the product is storable, will not purchase a title a second time and any later purchases will be of different titles. For such products the consumer enters the market in every period but will never repurchase the same variant previously purchased. The incentive to launch new variants will depend to a large extent on the costs of launch, but especially on the cross-price elasticity of demand between the new and the existing variants. It should be noted, however, that the market for existing variants may tend towards zero as time proceeds, and in such cases further innovation is a sine qua non for the producer to continue to exist. Of course, the impact of the introduction of a new variant on the revenue of the supplier of existing products will be internalized by a monopolist, but not by a new entrant. Again, this may provide different incentives to existing suppliers and to new entrants, even when it is counterfactuals that are being compared.

7.3.3 *Vertical product innovation*

Vertical product differentiation concerns products that come at different agreed quality levels. A useful example is aircraft travel, where there may be economy class, club class, and first class, with an agreed quality ordering (and, of course, with different prices). Vertical innovation will involve the introduction of a new product that is agreed by the market to be of different quality from existing products. The quality may be higher, and as discussed in earlier chapters this is, we believe, the scenario that the authors of the Oslo manual had in mind when defining product innovation. However, a lower quality product might also be introduced, and such an introduction is considered here to be an innovation too. We address here, in this context, the question of how much vertical product innovation will occur and what character such innovation will have.

The basic approach to modelling vertical product innovation, as developed by Shaked and Sutton (1982, 1983), presupposes a population identical in tastes but differing in income, which is, however, assumed to be uniformly distributed. For an initially simple approach, allow that there is only one product variant on the market and that that product is supplied by a monopolist. Assume that each potential buyer will buy only one unit per time period if s/he purchases at all, which buyers will do if they have a level of income above a threshold determined by the price of the product in that period (negatively) and by the quality of the product (positively). Given the distribution of income over the population,one may then generate a demand curve for the product (as a function of its own price).

Assuming that the costs of generating quality (R&D or design) increase with quality, it can be shown that, inter alia, a profit-maximizing monopolist supplier will choose a quality for the product that is lower when the costs of R&D or design increase more quickly with quality; higher quality variants cost more to produce when the density of buyers at any particular quality is less and when the highest income level is lower.

Dynamically one can observe what will happen as parameters change over time. In particular, if R&D or the adoption of new technology causes production costs to fall, then both the product price will be reduced and the product quality will be higher. Alternatively, if changes in the underlying technology cause reductions in the costs of generating quality, then again the quality of the offered product will be higher.

Although the single-good case is a simple one, it is not a very representative view of the world. We tend to observe that in most markets there are many goods of various qualities on offer that are marketed by a number of different suppliers. What are the drivers that produce this (vertical) product variety? Sutton (1986) argues that the crucial factors are the nature of production technology and consumer tastes, specifically, the relationship between consumers' willingness to pay for quality improvements and the increase in unit variable costs of production associated with such improvements.

To be explicit, Sutton allows that (i) all consumers have identical tastes but different incomes; (ii) a consumer's willingness to pay for quality improvements is an increasing function of income; (iii) a number of products of differing quality are offered on the market; and (iv) the cost of producing such products, and thus their price, increase with quality. Then, as detailed in chapter 6, if production costs increase steeply enough with quality, the consumers will partition themselves by income in such a way that brands of successively higher quality will be purchased by consumers in successively higher income bands. Fixed costs related to product variation will in fact determine the number of products offered to the market and thus the extent to which groups of consumers, as opposed to individual consumers, are serviced by a product. Therefore in a market with vertical product differentiation there may be, at any one time, a variety of products of different qualities that are sourced from one or more suppliers. The number of varieties will be such that the last product launched is just profitable and a further launch would not be profitable.

Over time, factors that would affect this number could well change, and this would lead to changes in either the number or the quality of products on the market. These might be for example changes in consumer income that affect demands for quality, changes in production costs through process innovation, and changes in knowledge and other factors that affect the costs of designing or generating new products.

When new products are introduced, these will, at least to some extent, take the market share of existing products. If the new products are launched by a monopolist, then the monopolist will have to bear all the cost of this market loss, in addition to launch and development costs. However, if the market is not monopolized, then the cost of lost market share from the later introduction of further new products may, to a large degree, fall upon other existing suppliers to the market. For a given gross revenue gain from innovation, this would suggest that more competitive markets would generate more vertical product innovation. In the literature relating to the impact of the number of suppliers on the extent of vertical variety, however, Greenstein and Ramey (1998) argue that, if the monopolist can be threatened with entry, monopoly provides strictly greater incentives to innovate. The emphasis upon entry conditions rather than differences in a predetermined market structure is particularly relevant when, as argued by Sutton (1986), it is more desirable to consider the supply industry structure as endogenous and determined by fixed costs. In this case the potential for entry is of particular relevance.

Lutz (1996) explores the impact of potential entry and concludes that, with quality-dependent costs identical with those of the potential entrant, the incumbent will always deter entry if possible (which is when fixed costs are high). If entry is deterred, quality will be set at a level lower than the optimal quality set if entry were accommodated. If entry is not blockaded, quality will be set at a level strictly lower than the optimal quality set under monopoly. Noh and Moschini (2006) also analyse the potential entry of a new product into a vertically differentiated market. They find that entry-quality decisions and entry-deterrence strategies are related to the fixed cost necessary for entry and to the degree of consumers' taste for quality. Such results suggest that potential entry may well impact upon quality and variety.

7.4 **Risk, uncertainty, and timing**

Innovation is an inherently risky activity. On the market side, the innovator is offering something new to the market and will not be able in advance to know for certain whether the market will welcome or reject that product (or how quickly it may be emulated). Just as there is market risk from product innovation, so there may also be risk in the development (technologically or creatively) of new products. There may be no certainty as to the outcomes of an R&D process or as to how new products would behave in the long run. At the time of writing, for example, Samsung has just withdrawn its Galaxy Note 7 phone from the market because of its tendency to catch fire. Product innovation failure rates of the order of 40 per cent seem to be indicated in the literature (Castellion and Markham, 2013). Such uncertainties may be of particular relevance when the innovator has to invest in production capacity in advance and also to spend on promotional activities, all of which will increase the required investment in bringing a new product to market.

Although, as Arrow (1962) argues, it may be socially more desirable if there were insurance markets on which risk could be traded (for then innovators could innovate and risk bearers could bear risk rather than the innovator having to do both), in general such markets do not exist. The three main reasons why the private sector may not provide insurance coverage and insurance markets (see a standard text such as Besanko and Braeutigam, 2015) may be absent are that (i) risks are highly correlated; (ii) adverse selection problems exist; or (iii) there are moral hazard problems. The first of these reasons is characterized by a situation where the losses insured against are positively correlated with one another (e.g. the losses may be cyclical) and private insurers may be unable to pool risk, and thus will not offer cover. The adverse selection scenario occurs when policyholders differ in their loss probabilities but insurance companies cannot distinguish high-risk individuals from low-risk individuals. As a consequence, many low-risk individuals may purchase little or no insurance coverage (because the presence of the high risks drives up the insurance premiums), thus potentially raising the overall riskiness of the insurance portfolio. Moral hazard occurs when the actions of the insured can affect the magnitude or the probability of a loss, but the insurance company cannot directly monitor the policyholders' actions. The optimal insurance contract will be a partial coverage contract, which provides the insured with some incentive to reduce expected losses but that does not expose the insured to large financial penalty if a loss occurs.

The main way in which risk can be shifted is by external funding of innovative activity. As we argue here and in chapter 8, such possibilities are limited, and in consequence the innovator himself will be the main risk bearer. The risks carried may well then affect the decision to undertake product innovation. A direct route by which this might happen is that such risks would increase the expected (or mean) rate of return from a new product and would thus lead to less innovative activity. On the other hand, one could also argue (see Brealey, Myers, and Allen, 2014, and PA Consulting Ltd, 2014) that, because risk may be reduced by diversification, firms would undertake more product innovation in order to offset the risk of loss on one product through poor market reception with the potential gain from good market reception on another. Thus, for example in the face or market uncertainty, a firm may launch several (vertically or horizontally) differentiated versions of a new product in order to spread the risk that some would fail. Using a similar line of argument, in the face of technological uncertainty, in order to spread risk, firms may pursue a wider set of projects in their R&D or creative portfolio. The extent to which such outcomes arise will largely depend on the cost of pursuing these strategies. In certain cases, for example in a number of creative industries (books, recorded music), the costs of product launch to the innovator are low; it should be of little surprise, therefore, that so many products are launched (see chapter 3), most of which fail.

Uncertainty may also impact upon the timing of innovation. The choice of the timing of an innovation is a trade-off between the higher cost of developing a new product early and the market benefits of being early to the market. Although there is some argument that early innovators do not always benefit most (Lanzolla et al., 2010), the argument is best illustrated by considering that early appearance on the market yields a greater return. With certainty, then, one may state that the ideal launch date for a new product would be the date when the discounted value of the extra (marginal) cost of development that would have to be incurred to bring the date forward is equal to the discounted extra (marginal) profit to be gained from doing so. In the presence of uncertainty, the costs and benefits would have to be discounted at a higher rate. Traditionally it is argued that the launch date would then be the first date at which the discounted expected benefits exceed the discounted expected costs of development.

The theory of real options (Dixit and Pindyck, 1994) suggests, however, a different criterion for determining the launch date. This theory allows that decision makers learn over time. As they learn, their evaluation of risks and uncertainty are modified (reduced). They are also aware that they will learn over time. Thus decision makers may delay decisions upon launch dates in order to gather more information and be more certain about the correctness of their decisions. The launch date will then be determined not by the first date at which the discounted expected benefits exceed the discounted expected costs of development, but by that date when (i) the expected discounted benefits exceed the cost of development and (ii) the learning process is not expected to offer information of greater value than the loss resulting from further delay.

7.5 **Constraints to innovation**

Having discussed the incentives to undertake product innovation, it is also relevant to consider the reverse, that is, the barriers to innovation. There may be many such barriers, in fact some of them may just be the opposite side of the coin to the incentives because, in a competitive market, the incentive to one firm may just act as, or lead to, a barrier or disincentive to other firms. In order to identify the most important barriers, we rely upon survey data from the community innovation survey (CIS). These data are presented and discussed in chapter 8, where we analyse data from the UK 2015 Innovation Survey and from the 2012 CIS concerning self-reported barriers to innovation in seventeen EU countries. The data are not restricted to product innovators but refer to firms that are innovative in some way (not necessarily in product innovation), which, it is argued in chapter 8, is appropriate in the circumstances.

Four main classes of barriers are identified: costs, knowledge, market factors, and other factors. From this list, four most important barriers are identified: direct innovation costs that are too high; a lack of qualified personnel; a lack of finance at an appropriate cost; and an excess of perceived risk. It would appear, however, that the last two are so closely related (in that the source of finance determines the risk carried by the innovator) that they cannot be discussed separately. Although product market issues (competition under various guises) and regulations are also identified as constraints, we do not consider the former as a constraint, it is more to do with the market environment, and the latter encompasses a very wide field, which would be inappropriate to discuss here but is addressed in chapter 13 (on policy). There are thus three main barriers to innovation to be discussed: innovation costs; risk and finance; and the availability of qualified labour.

7.5.1 *Costs of innovation*

The fact that the costs of innovation impact upon innovative behaviour is obvious and is a mainstay of all economic analysis of the issue. If costs are higher, expected returns will be lower and the amount of innovation undertaken will be less. It is worth, however, mentioning two refinements of the argument. On the grounds that speed to market may be a prime determinant of costs, a cost constraint may in fact reflect a situation where the required speed to market for success in the face of competition is too fast. The second point is that a cost constraint may in fact reflect a situation where a too high level of costs of innovation could indicate that the required expenditure to innovate is large not just relative to expected returns but relative to the resources available to the innovator—that is, the greater the cost, the more it may be that the potential innovator will have to 'bet the farm' on the intended innovation and will be unwilling to do so. It is noticeable in the data that costs are considered to be a more important constraint for the smaller than for the larger firms, and hence there may be some element of this unwillingness or reluctance in the data.

7.5.2 *The availability of qualified labour*

Qualified labour is an obvious requirement for innovative activity, be it in the design or in the development of new products, in establishing the means to produce such products, or in managing organizations in a successful way. The availability of qualified labour is a key element in a National System of Innovation (Nelson, 1993) and will to large degree reflect national (regional, local) investments in education and training. In fact the availability of staff may be a prime determinant of the location of firms' innovative activity.

We note that there is wide dispersion across countries in the extent to which such availability is an important constraint on innovation. In a perverse way, however, it may be the most active innovators who suffer most from a shortage of qualified staff. Those who do not innovate will have less need of such staff.

However, the availability of qualified labour is not only a national issue. It is also a matter for individual firms and innovators, who will train in house and enable their staff to learn and gain expertise. But qualified staff may choose to change employer (see, for example, Rao and Drazin, 2002). Then the argument is that firms that undertake training may bear the cost of training, whereas poachers may attract skilled labour by offering higher wages and do not have to bear such costs. The incentives to train may thus be distorted. For these reasons education and training are often argued to be a social rather than a corporate responsibility.

7.5.3 *Risk and finance*

The financing of innovation (in terms of either availability or cost) is, along with risk, one of the main constraints on innovation in the United Kingdom according to the data; and it is also of great importance in the other EU economies surveyed. All innovation, be it product or process innovation, has the character of an investment, that is, funds are expended in advance of the product's launch and are returned only if the product is successful. The costs involved may encompass R&D, design costs, launch costs, various fees, and many similar expenditures. These expenditures need to be funded.

Funding of these investments may be generated from sources internal to the innovating firm (from the innovators' own funds) or from external sources (such as bank borrowing, equity, loans, trade credit, etc.). A basic difference between internal and external funds is that, in the presence of uncertainty, the use of external funds shifts some of the risk of the investment to the funder (partly of course at the cost of a higher repayment required later), whereas internal funding will mean that the innovator bears the risk. For this reason risk and finance are considered here together rather than as separate issues. In fact it may only be through the use of external financing that an innovator gets insured against failure (at least up to the funds raised externally). However, insurance markets tend to be imperfect, or not available, because (i) risks are highly correlated; (ii) adverse selection problems exist; or (iii) there are moral hazard problems.

The 2015 UK Innovation Survey indicates that financial constraints do exist in the sense that respondents to the survey consider them a barrier to innovation. Canepa and Stoneman (2007) have approached this issue using responses to the UK CIS 2 and 3. Analysis of the CIS2 data (individual returns for UK firms, 1994–6) indicates that, correcting for firm size, there is evidence that a firm in a high-tech sector has more chance of experiencing a greater impact from financial constraints than a firm in a low-tech sector. Further results using the CIS2 data also provides clear evidence that, once one has corrected for industrial sector, small firms are at greater risks of experiencing a strong impact from financial constraints than do large firms. The CIS3 data set (individual returns for UK firms, 1998–2000) confirms these results. Overall, one may thus conclude from CIS survey data that financial factors do impact upon innovative activity. That impact is more severe in higher tech sectors and for smaller firms. However, Mina et al. (2013) use alternative survey data that take a more nuanced approach to look at the relationship between the demand and supply of finance for innovation and find less support for the existence of such constraints.

There is also a wider literature, beyond the CIS and other surveys, which starts with the work of Stiglitz and Weiss (1981), who consider a firm to be credit-rationed if it does not get as much credit as it wants, although it is willing to meet the conditions set by the lender on equivalent credit contracts. According to Hall (2002), a financial constraint is said to exist when, even if there are no externalities involved in the firm's investment activity, there is a wedge between the rate of return required by an entrepreneur investing his/her own funds and that required by external investors. In essence, therefore, a firm is considered credit-constrained or financially constrained if it cannot raise external funding at the market price or if, in order to raise external funding, it has to pay over the market price.

The theoretical groundings for propositions related to the existence of financial constraints (see, for example, Bond et al., 2003 and Hubbard, 1998) are primarily based on there being asymmetric information between firms and the suppliers of finance with associated moral hazard and adverse selection problems (these are the same factors that prevent the existence of a full set of insurance markets). Small firms may then be more prone to financial constraints, as they offer riskier investments (with innovation of a type less likely to have been undertaken elsewhere, and thus with particular problems in observing systematic risk), greater information asymmetry, shorter track records, less collateral, and assets that are less realizable.

Most of the empirical literature that follows this tradition (for example, Bond et al., 2003) has then tested whether the firms' investment in plant and machinery is affected by financial and liquidity constraints by exploring whether such investment is particularly sensitive to cash flow—on the grounds that a changed availability of internal funds would not much affect investment if external funds were not constrained. However, of greater relevance is a more limited literature (surveyed by Hall, 2002, and by Hall and Lerner, 2010), which in a similar way looks at the impact of financial constraints upon R&D expenditure as an indicator of innovation. Recently Hall et al. (2016) have returned to this issue using a variety of methodologies to confirm that financing constraints, whether measured as cash flow or by firm survey responses, do affect negatively the level of R&D investment chosen by European firms, especially if they are more technology-intensive or smaller (or both)—a result that confirms that the evidence for a financing gap for large and established firms is harder to establish (e.g. Bond et al., 1999). In general, the results underline the fact that in the EU ten countries, innovative firms may have problems of credit access, especially in an environment characterized by macroeconomic recession and uncertainty. However, they also contain hints that higher quality firms—proxied by higher total factor productivity (TFP), exporting, and being more technology-oriented—are able to maintain their activities and productivity in the presence of economic downturns.

7.6 Conclusions

The aim of this chapter has been to explore the incentives that drive firms to undertake product innovation, given the nature of the markets in which they operate. This has encompassed a discussion of the driving forces that encourage product innovation, for example innovation by others or the ageing of an existing product line; but the main incentive is considered to be the search for profits. The chapter has also attempted to address decisions related to the extent of innovation in general, especially by addressing horizontal and vertical product innovations separately and the location of innovations in product space. Woven into the analysis are discussions of the role of market structures in the product innovation decision, the issues raised by uncertainty in the innovating environment, and also issues related to emulation and copying.[1] We have also explored those constraints to product innovation that, as survey data indicate (see chapter 8), are most important, namely innovation costs, risk and finance, and the availability of qualified labour. In chapter 8 we attempt to put some further empirical flesh upon the conceptual bones discussed here.

[1] Although this debate has been wide-ranging, only by considerable hand-waving could it be said to provide any foundation for the patterns of market dynamics that are the supposed result of market maturity, as discussed in chapter 4.

8 Empirical Evidence on the Determination of the Extent of Product Innovation

8.1 Introduction

The objective of this chapter is to consider the empirical evidence available to us on the determinants of product innovation across and within firms. Having explored the theoretical or conceptual groundings of product innovation in chapters 7, this is a natural next step. The chapter obviously also links back to chapters 2, 3, and 4, which deal with the definition, sources, and extent of product innovation and where we have illustrated differences in rates of product innovation across sectors and firm sizes. This chapter is not particularly traditional in its approach. A traditional approach would usually consider the relationship between the extent of product innovation and a number of explanatory variables suggested by the theoretical literature and would use hypothesis testing to see whether any of the hypotheses is valid or can be refuted. The problem with such an approach is that, as explored in chapter 4, there are difficulties in measuring product innovation with precision, both in its frequency and intensity. One may explore data on R&D, patents, copyrights, and new product launch counts, for example; but each measure, although having some advantages, also has limitations as an indicator of product innovation.

As an alternative, we observe that one notable aspect of innovative behaviour that is revealed in survey data on innovation is that firms that are innovative in one dimension are also innovative in others. Thus, for example, firms that are product-innovative are often also process-innovative, innovative in marketing, and innovative in organizational characteristics. This leads us to argue that, in essence, the main determinant of whether a firm is a product innovator is whether the firm is an innovative firm or not an innovative firm. Taking this one step further, to explore what determines whether (or to what extent) a firm will undertake product innovation is to explore whether (or to what extent) a firm is or is not an innovative firm.

Pursuing this argument, we proceed by first exploring survey data on the innovative activity of firms to illustrate how product innovation and other forms of innovation are linked and, using principal components analysis, we group firms into clusters of innovative and non-innovative firms. We are then able to explore the characteristics of the firms in different clusters. Building upon the finding that the main determinants of whether a firm undertakes product innovation is whether that firm is an innovative firm, we then turn to the wide literature on the characteristics of innovative firms to fill in more details on the determinants of product innovation in firms. Whereas survey data tend to indicate whether firms are innovative or not in certain directions, the wider literature presents the advantage that it more often concentrates on the extent of innovation (e.g. the number of patents issued, or the extent of R&D spending). Finally we return to the survey data to provide more detail about the characteristics of innovative firms by considering survey responses related to the constraints that firms consider to have limited their innovative activity.

8.2 Survey data

Battisti and Stoneman (2010) use data from the fourth community innovation survey in the United Kingdom (CIS4) to explore a range of innovative activities—encompassing process, product, machinery, marketing, organization, management, and strategic innovation—across 16,383 British companies in 2004. Using data on whether or not the firms in the sample had undertaken each of the several innovative activities, the first finding is that, Kendall's tau-b correlation coefficient (a non-parametric measure of association based on the number of concordances and discordances in paired observations), for all the seven innovation variables, indicates that the pair-wise degree of association is significantly different from zero. These results indicate that undertaking one innovative practice is not independent of undertaking another innovative practice, and hence that the adoption of any practice is correlated with the adoption of all others. However, the degree of association differs in intensity and varies from innovation to innovation. For illustration, it is found that, out of the firms that undertake product innovation, 43 per cent also undertake process innovation, 32 per cent acquire new machinery, 34 per cent undertake marketing innovation, 27 per cent undertake organizational innovation, 21 per cent undertake management innovation, and 27 per cent undertake strategic innovation.

The initial analysis shows that, even after taking account of potential common driving factors, there is a significant and positive degree of association between all pairs of innovations undertaken, which indicates complementarity in the simultaneous undertaking of the different innovation listed, although the degree varies and there are differences in intensity from pair to pair of innovations. This suggests that there exist important synergies generated by joint innovation activities, although some innovations are more influential and versatile than others. This is a finding also replicated in other work, for example Bartoloni and Baussola (2015), Frenz and Lambert (2008, 2009), and Cozzarin (2016). The implication is that to concentrate on the analysis of the adoption of single innovations in isolation would be misleading; it is far preferable to consider the joint undertaking or adoption of complementary innovations.

To pursue this approach, Battisti and Stoneman (2010) use iterated principal factor analysis (IPFA) in order to identify the underlying pattern of intensity of use in different innovative practices by the firms. IPFA models the correlations among innovation activities and allows one to summarize the heterogeneity of use of a set of seven innovations via a reduced number of latent factors. These factors are intended to pick up those underlying pattern of use that can explain the largest proportion of variability of the joint adoptions, and so identify the innovative practices that play the major roles in the overall innovative activities of the firm. Two factors are identified: the first factor (factor 1) accounts for 83.5 per cent of the total variability in firms' innovative activity and is driven by the extent of use of strategy, management, organizational, and marketing innovations, here jointly labelled organizational innovations. The second factor (factor 2) explains 16.5 per cent of the remaining variability in the heterogeneity of use of innovative activities by the firms in the sample and is driven by product, process, and machinery innovations, which are generally labelled technological innovations. Interestingly, it appears that most of the heterogeneity of innovation patterns across firms is largely related to organizational variables, that is, although the innovation literature has been mainly concerned with 'traditional' or technological innovations, 'wider' or organizational innovations play a dominant role in the innovative activity of UK firms.

Once the two factors have been identified, a cluster analysis is performed that identifies three clusters of firms in the sample on the basis of the intensity of use of the seven innovations. Cluster 1 (9,317 firms, or 57 per cent of the sample) contains the least 'innovative' firms. Within this cluster less than 2 per cent of the firms report having carried out organizational innovative activities (in fact undertaking organizational innovations at levels below the sample average), although about 22 per cent have introduced technological innovations. Only 6 per cent of these firms have developed new products, and only 2.3 per cent of those products were new to the market rather than just new to the firm. The firms in cluster 2 (3,881 firms, 23.6 per cent) and in cluster 3 (3,185 firms, 19.4 per cent) use organizational innovations progressively more intensively; the use of technological innovations also increases as one moves from cluster 1 through to cluster 3, the intensity of all innovation activities being highest in cluster 3, where the majority of firms have undertaken each of the seven innovations. Given that cluster 1 has the largest number of firms and cluster 3 has the smallest, these data suggest that about

19.4 per cent of UK firms operate well above average in terms of innovative activity, while 56.9 per cent perform below the average.

Interestingly, across the clusters it is found that factor 1 innovation is positively associated with factor 2 innovation, which suggests that organizational innovations and technological innovations do not represent substitute, alternative, or competing innovation strategies but rather are complements with positive synergistic effects. If the factors had been substitutes, one would expect to have seen some firms using organizational innovations intensively but not technological innovations and other firms using technological innovations intensively—but not organizational innovations. As such patterns are not observed, one may reliably conclude that organizational and technological innovations are complementary.

In a further paper, Battisti and Stoneman (2013) perform a similar exercise upon data from an enterprise-level survey of innovation activities of firms in the United Kingdom undertaken for the National Endowment for the Science Technology and the Arts (NESTA; see Roper et al., 2009, for details). An important empirical divergence between this paper and earlier versions is that the data used have a very obvious and deliberate bias, away from the manufacturing sector and towards services. Six of the nine industries in the chosen sample may be considered service industries and contribute 77 per cent of the sample firms. Small firms (5–19 employees) represent 49 per cent of the sample and 20 per cent are large firms (100+ employees), the balance being medium-size firms.

The main value added that comes from this study is that it separately identifies the sourcing of new ideas and the introduction of new ideas. CIS data concentrate upon the undertaking of different innovations and say little about the sourcing of such innovations. It is, however, expenditure on sourcing that has often dominated the empirical literature on the determinants of innovation, particularly expenditure on R&D. It is thus on the results about sourcing that we concentrate here. We note, however, that the results about the introduction of new ideas are little different from those based on the CIS data.

In Battisti and Stoneman (2013), five sourcing activities are defined and measured, each of which reflects a means by which firms may acquire technological knowledge: engaging in activities related to R&D; engaging in activities related to design; engaging in changes to business processes; sourcing new equipment and software; and new branding and marketing activity. Note that what is being measured is whether firms engage in the activity or not, and not the intensity of their engagement in each activity. The most frequently observed activities concern the sourcing of new software and the branding and marketing activity: half of the firms in the sample have been active in both. But only 21 per cent of the sample reported having engaged in R&D activities. The proportion that undertakes R&D is lowest in industries such as accountancy (6 per cent) and legal services (10 per cent). Firms in accountancy and legal services are more active in software acquisition and branding. This is unsurprising, for research in services is not carried out in formal R&D labs, as it might be in traditional manufacturing. It is also clear from the data—and equally unsurprising—that there is a significant positive association between the firms' propensity to source and the firms' propensity to use new ideas.

Initial analysis of the sourcing of different innovation across firms revealed that calculated tetrachoric correlation coefficients in the sourcing of pairs of technologies are all positive and significantly different from zero. The data indicate, for example, that, out of all the firms that do R&D, 54 per cent also do design and, out of all firms that source new software, 35 per cent also engage in new business processes. Thus firms tend simultaneously to source a number of new ideas. The adopted methodology based upon the decomposition of the tetrachoric correlation matrix led to the identification of just one factor in which the weight on design is greatest; R&D and sourcing new business processes carry a slightly lower weight, while software and branding carry the lowest weight. Having just one factor allows a simple calculation for each firm of an aggregated summary measure of overall sourcing: the factor score.

Decomposing the scores of the firms in the sample using the criteria of distance minimization, a two-step cluster analysis identifies three clusters of firms (A, B, and C) in which firms share similar patterns of intensity of sourcing. The intensity of engagement in all activities steadily increases from cluster A to cluster C, which suggests that the firms that are the most active in sourcing new ideas of one type are also the most active in sourcing other types. These observations reinforce the view that different types of ideas complement one another rather than being substitutes for one another. The first sourcing cluster, labelled A, contains 44 per cent of the sample. These firms have a level of engagement in sourcing new ideas that is low. Very few undertake R&D, or design activities, or seek changes in business processes, but about a fifth of them source new equipment or software and/or explore changes to branding or marketing. Cluster B contains firms that engage only moderately in sourcing new ideas, although considerably more extensively than firms in cluster A (especially as regards R&D or design), while cluster C (15 per cent of the sample) contains firms that engage most in sourcing, with an intensity almost six times higher than that of firms in cluster B.

8.3 The determinants of innovativeness

Having indicated that the firms' propensity to source or innovate in one direction (say, R&D, or introduction of a new product) is closely associated with their propensity to source or innovate in another (say, undertake design activity or introduce a new process), we argue that the most relevant way to explore the determinants of innovative activity is to consider the differences between innovative and non-innovative firms rather than attempt to analyse one indicator (in the case of the greatest importance here, the extent of product innovation). This also implies that various literatures that use different, but usually single, indicators of innovative activity (e.g. patenting activity, R&D spending, spending on design, new product or process introductions) should all be telling the same story and thus should all be of relevance in helping to answer the question of what determines the extent of product innovation.

There is a large (and still growing) literature on what, theoretically, may be expected to affect the innovative behaviour of firms, and a similarly large empirical literature testing the many hypotheses proposed. We do not intend here to undertake a detailed survey of these two literatures. There are several such surveys already in the public domain. Among the most informative are Cohen's (1995, 2010) surveys, while Becker (2013), although concentrating upon the determinants of R&D, is also informative. These sources also offer very detailed bibliographies; therefore referencing is kept to a minimum here.

Much of the literature in this field has its origins in the seminal work of Schumpeter (1934), which has led to what has become known as 'the Schumpeterian hypotheses'. In their simplest form, these essentially state that innovativeness increases as the degree of competition in the industry to which the firm belongs diminishes, and that innovativeness is also positively related to firm size. The key aspects of the logic that underlies the first proposition are the following: (i) if a firm has monopoly power on its current product(s), it may be able to extend this power to its new products, for example through its leadership over distribution channels; (ii) a monopoly firm, through higher profit levels, may be able to finance innovation internally, which is an advantage per se and limits the necessary disclosure of information concerning the innovation that borrowing involves and from which potential rivals may take advantage; (iii) finally, a firm with monopoly profits may well be able to hire more or better skilled workers and scientific personnel. The main arguments that support the suggested positive impact of firm size can be similarly listed as being about better access to capital, skilled labour, ability to deter copying, and also—to the extent that size reflects past success—size: firm size may encourage innovation via a 'success breads success' hypothesis.

Even so, many small firms are innovative, although the pattern is heterogeneous; for example de Jong and Vermeulen (2006), drawing upon a database of 1,250 small firms across seven industries, reveal (1) some major differences in the extent to which small firms use innovative practices and (2) the connection of new product introductions with the use of innovative practices.

Audretsch et al. (2016), however, on the basis of a rich data set of entrepreneurs receiving research funding through the US Small Business Innovation Research (SBIR) program, argue that business-based human capital and prior business experience are not correlated with innovative activity in small firms, whereas academic-based human capital is positively correlated. Klepper (1996) argues as well that small firms may find it easier to innovate in the early stage of the life of an industry because economies of scale at this stage are less important and firm mobility is high by comparison to later periods in the cycle.

Over time, the stress upon the direct importance of these two factors (monopoly power and firm size) has waned, more emphasis being placed on technological factors that may differ considerably across industries, market factors (e.g. market size or geography), firm characteristics other than size (e.g. access to finance and appropriability conditions), government policy measures such as R&D, and local or international spillovers. Our personal reading of the large literature, based on and echoing the findings of the surveys of Becker (2013) and Cohen (1995, 2010) suggests that the empirical findings as to the determinants of innovativeness indicate as follows:

(i) Technological characteristics of industries matter, some offering more or cheaper opportunities to innovate and others being less conducive. The nature of innovation also differs across industries, as the emphasis on different types of innovation (such as product, process or organizational innovations) varies.

(ii) There is an increasing realization that innovativeness and firm and market characteristics are not linearly related but rather that there are many and various feedback loops between them.

(iii) Generally, the two most important firm characteristics affecting innovativeness are internal finance, most notably cash flow, and sales, both being positively correlated with innovative activity. Stronger sensitivity is found for small firms and young firms, and perhaps for firms in the United States and United Kingdom rather than in continental Europe (perhaps a reflection of different financial systems). The constraints may also differ between different types of R&D (e.g. basic and applied), but the evidence here is scarce.

(iv) A very large literature relating innovation to rivalry or competitive pressure is coalescing around the view that, if the former is plotted against the latter, then the resultant curve would be of an inverted U shape, the precise shape and position of that curve also depending on firm sizes and the costs of innovation in the industry.

(v) Competition in foreign markets, via exporting, is predominantly found to have positive effects on R&D and innovation at home.

(vi) There is wide support for a variety of spillover effects, for example among firms, between universities and firms, and also across national boundaries.

(vii) Evidence suggests that R&D tax credits and direct state subsidies have positive effects on private R&D investment, although there is some indication that the effect of a subsidy may be non-linear.

Because most of the literature encompasses either R&D spending or process innovation as opposed to product innovation itself, it is worth referring explicitly to three papers that consider product innovation explicitly. Fritsch and Meschede (2001) estimate the firm size elasticities for process and product R&D in a sample of German manufacturing firms, suggesting that, although the size effect is stronger for process R&D, the difference from the impact on product R&D is slight. Brouwer and Kleinknecht (1996) find that, typically, larger companies are more likely to innovate but smaller firms, once they have adopted an innovation, are more innovative than larger ones in terms of the extent of sales due to new products (except in the case of small manufacturing businesses introducing product innovations new to the firm). Corsino et al. (2011) estimate the impact of R&D expenditure and firm size on new product announcements within a sample of leading companies in the semiconductor industry. Although the impact of R&D and firm size is positive, their estimates suggest that, in any case, the impact illustrates decreasing returns.

To supplement the summary of the empirical findings in the literature just presented with a view more related to the concept of the innovative firm, we turn to the results in Battisti and Stoneman (2010, 2013). In the first of these papers three clusters of firms (1, 2, and 3) were identified on the basis of their overall innovative activity (1 being least innovative and 3 being most innovative). Although it is dangerous to infer causality from a single cross-section, one may identify certain characteristics of firms in the different clusters that could suggest a link to innovativeness.

First, an analysis of the distribution of firms across industrial sectors by clusters indicated that in every sector cluster 1 contained the largest number of firms, suggesting that the distribution of firm innovativeness is skewed. Second, firms operating in the service sector are no more likely to belong to cluster 3 than firms in other sectors. Third, within the production sector, perhaps unsurprisingly, firms in mining and quarrying, electricity, gas and water supply, and construction are the least intensive innovators. By contrast, firms in high-technology sectors such as the manufacturing of electrical and optical equipment, the manufacturing of transport equipment, and the manufacturing of fuels, chemicals, plastic metals, and minerals are more intensive innovators. The two sectors with the highest percentage of low-intensity users were in services (retail trade and hotels and restaurants).

Battisti and Stoneman (2010) also find the following results. (i) The extent of firm innovativeness seems to increase with firm size, cluster 1 firms being mostly small, cluster 2 firms being mainly of medium size, and cluster 3 firms being medium to large (however, the standard deviations are very large, suggesting that averages can be highly misrepresentative). (ii) The proportion of establishments that carry out in-house R&D is lowest in cluster 1 and highest in cluster 3, reflecting an hypothesis that formalized R&D exerts a positive impact upon innovation activity. (iii) The proportion of employees with a degree in science, engineering, or other subjects increases progressively as one moves from cluster 1 to cluster 3, confirming the importance of the link between innovation and skills. (iv) The percentage of firms that received public support increases with the extent of innovative activity carried out by the firm, reaching a peak of 25 per cent in the highly innovative group (cluster 3). (v) The proportion of firms that are part of a group (vs. independent establishments) is higher in cluster 3 than in the other clusters. However, no significant differences across clusters were found with respect to the age of establishments or whether the market for the firm's final product was international.

It must be stressed, however, that it is not possible with these data to consider cause and effect. Thus, for example, one is unable to say whether firms are large because they are innovative or innovative because they are large. Similar statements can be made with respect to spending on R&D, employment of graduates, and receipt of public support. One is thus unable to say whether only 19.4 per cent of UK firms operate well above average in terms of innovative activity while 56.9 per cent perform below the average because of their character or their character is what it is precisely because they underperform.

Battisti and Stoneman (2013) also find three clusters of firms in their sample: cluster A, which sources innovation least intensively, cluster C, which sources innovation most intensively, and cluster B, which is intermediate between A and B. The authors' analysis of the distribution of firms across the three sourcing clusters provides support for the hypothesis that the levels of the sourcing of new ideas by firms differ across firm and industry characteristics. There are differences across industries that reflect perhaps technological characteristics of products and market structure; for example, a large proportion of the firms in architecture and the software industries are in the highly active cluster C, whereas more than 50 per cent of firms in accountancy and construction are in the less active cluster, A. However, the patterns for automotive, construction, and energy production do not differ markedly from patterns in other industries; thus it is concluded that manufacturing firms are no more or less active in sourcing new ideas than firms in services.

Looking at the characteristics of firms in the three different sourcing clusters, the authors found that, although there are no significant differences with respect to the age of the business, a firm is more likely to be a member of the high-sourcing cluster C if it is larger, is foreign-owned, offers both physical products and services, has a proportion of employees with a degree between 11 per cent and 20 per cent, has a higher percentage of turnover from exports, and has some product uniqueness. On the other hand a firm is more likely to be found in the low-sourcing cluster A if it is UK-owned, offers only products or services alone, has less than 5 per cent of employees with a degree, and does not export. Once again, a warning about causality is in place. It is clear, however, that evidence based on the characteristics of the innovative firm supports the findings that use more traditional indicators.

8.4 Constraints on innovation

Finally, having discussed the existing literature on empirical studies of innovative activity and placed some emphasis upon the character of the innovative firm, we add further insight by considering barriers to innovation. This will become especially relevant when policy is discussed, in chapter 12. There may be many such barriers (in fact some of them may just be the opposite side of the coin to the incentives to innovate, in that, in a competitive market, what is an incentive for one firm may just act as or lead to a barrier or disincentive for other firms). To initiate the discussion we provide some data on barriers to innovation from the UK 2015 Innovation Survey, as detailed in Table 8.1. The data are not restricted to product innovators but refer to firms that are broader innovators, that is, innovative in some way (not necessarily in product innovation).[1] This is in line with our findings about the concept of the innovative firm (a firm innovative in one direction is innovative in others).

The data are organized to cover four main classes of barriers: costs, knowledge, market factors, and other factors (regulations). The data clearly indicate that the major perceived potential barriers to innovation are cost factors, although lack of qualified personnel, regulations, and (product) market factors also have some role to play. However, the market factors include uncertain demand, which could be lumped with excessive perceived economic risks (a cost factor); and it would also seem fitting to consider the lack of information on markets and technology as another aspect of risk. Thus in our view the list of the most important barriers consists of direct innovation costs that

[1] More precisely defined as any firm that has undertaken any one or more of the following: introduction of a new or significantly improved product (good or service) or process; engagement in innovation projects not yet completed or abandoned; new and significantly improved forms of organization, business structures, or practices and marketing concepts or strategies; or investment activities in areas such as internal R&D, training, acquisition of external knowledge, or machinery and equipment linked to innovation activities.

Table 8.1 Percentage of broader innovators, self-reported perception of potential barriers to innovation, UK Innovation Survey 2015

Size of enterprise (employees)	10–250	250+	All
Cost factors			
Availability of finance	17	9	17
Direct innovations costs too high	15	11	14
Excessive perceived economic risks	14	9	14
Cost of finance	14	8	14
Knowledge factors			
Lack of qualified personnel	8	6	8
Lack of information on markets	3	2	3
Lack of information on technology	3	3	3
Market factors			
Market dominated by established business	10	6	10
Uncertain demand for innovative goods	8	7	8
Other factors			
UK government regulations	7	6	7
EU regulations	6	6	6

Source: DBIS (2016a).

are too high; a lack of qualified personnel; a lack of finance at an appropriate cost; and excess perceived risk. It would appear, however, that the last two are so closely related—in that the source of finance determines the risk carried by the innovator—that they cannot be discussed separately. Market dominance by others (the extent of competition in the market) and regulations are also relevant.

A further relevant set of empirical observations comes from CIS 2012 and relate to self-reported barriers to innovation in seventeen EU countries.[2] The sample includes both innovators and non-innovators (in this it differs from the UK sample, which only refers to broadly innovative firms) and has a slightly different list of barriers from those listed in the UK survey. The entries in Table 8.2 indicate the range of responses across all seventeen countries and the proportion of firms that considered each of the factors to be a highly important barrier to innovation. The most obvious result from these data is the high degree of heterogeneity in responses across countries. Some factors considered of great importance in one country are considered of little relevance in others. This is a clear reflection of different systems of innovations (Nelson, 1993) in different countries. However, taking a high-level overview of the data, the issues that we considered to be of main importance in the UK

[2] Bulgaria, Germany, Estonia, Greece, Croatia, Italy, Cyprus, Lithuania, Hungary, Malta, Netherlands, Austria, Poland, Slovenia, Sweden, Serbia, and Turkey.

Table 8.2 Obstacles to innovation considered highly important, percentage of respondents (innovators and non innovators, seventeen EU respondent countries), 2012

Obstacle	Minimum	Maximum
High cost of access to new markets	7.8	41.4
Lack of adequate finance	7.3	39.4
Lack of qualified personnel	5.1	28.1
Dominant market share of competitors	11.4	28.5
Innovation introduced by competitors	3.8	31.3
Lack of demand	12.3	39.3
Strong price competition	8.0	63.1
Strong competition on product markets	13.6	43.0
High cost of meeting regulations	6.0	42.6

Source: Eurostat (2016), *Obstacles to Innovation Considered Highly Important by Innovative and Non-Innovative Enterprises*, European Commission Database, March; own calculations, found via Eurostat data explorer at appso.eurostat.ec.europa.eu/nui/show.do? data set=inn cis8 obst&lang=en.

case may also be considered to be so in at least some other EU countries, that is, innovations costs, risk and finance, and the availability of qualified labour. Product market issues and regulations are also relevant.

8.5 Entry, start-ups, foreign direct investment, and imports

Although survey data yield considerable insight into the innovative activity of domestic companies, there are three sources of product innovations that are not reflected in such exercises, primarily because the surveys mainly reflect product innovation activity in domestically based production units. The first of the omitted sources is new firm entry or the establishment of new start-up companies, which is omitted because these lie outside the sampling frame of most innovation surveys. Secondly, surveys will only to a limited extent pick up the creation of new production facilities in the domestic economy by overseas-based companies. Thirdly, innovation surveys will not reflect product innovation based on imports to the domestic economy from overseas producers. A full picture of the determinants of product innovation in an economy thus requires some insight into these issues.

Whether new entry means more innovation overall is by no means clear. For example, although less entry means fewer new products launched by entrants, it might mean that existing firms would have more profits to invest and may face less competition, both of which may cause them to launch more innovations (although this is by no means certain).

Although the literature emphasizes that monopoly power and firm size may, singly or together, enable firms to deter new entry, there are also strong arguments in the literature that new firm entry into markets varies with the stage at which the industry is in its life cycle. Mueller and Tilton (1969) indicate four basic stages of technological growth for an industry: innovation, imitation, technological competition, and standardization. Uncertainty is the main barrier to entry at the innovation stage, but this rapidly declines, and at the imitation stage entry may be more rapid, as R&D costs are relatively low. In the last two stages, which are longer than the previous two, entry is slower. Under technological competition, R&D costs are high and act as a barrier to entry. Under standardization, as time proceeds, the declining level of sales and the lower expected profits deter firms from entering.

Taking an evolutionary perspective, Gort and Klepper (1982) specify five stages of the market for a new product: (i) the first stage is the commercial introduction of a new product by its first producer; (ii) the second stage is a period of sharp increase in the number of producers; (iii) in the third stage there is a sort of stabilization, inflow and the outflow of new firms being reciprocally balanced; (iv) the fourth stage is a period of negative entry; and (v) in the fifth stage there is zero entry. It may also be argued that entry differs across the stages because it depends on access to relevant and transferable knowledge and, in the early stages, the major sources of information are external to existing firms and accessible to potential entrants, whereas in the later stages the sources are internal (e.g. learning by doing) and not accessible to potential entrants.

There is a closely related and extensive literature upon the determinants of start-up activity, but we have found that the survey by Cincera and Galgau (2005) provides useful insight. They classify the factors that influence entry and exit decisions of the firms into three categories: (i) firm-specific; (ii) industry-specific; and (iii) country-specific. The first category includes firm size, age, growth rate, excess capacity, financial structure, and managerial turnover (the signs of impact are obvious in most cases). The second category includes limit or predatory pricing behaviour, lagged entry and exit rates, minimum efficient scale of production, capital intensity, resource intensity, the degree of maturity of the industry, concentration, and the degree of differentiation. Finally, the third category includes degree of economic development, frequency and size of macroeconomic shocks, access to start-up capital, (past) profits, adjustment costs, subsequent growth rates of survivors, advertising intensity, R&D intensity, and innovation.

From this set of potential factors, Cincera and Galgau (2005) initially conclude, on the basis of both theoretical and empirical literatures, that in general new start-ups have a low survival rate. They also conclude on determinants that (i) there is some evidence that higher levels of start-ups encourage further start-ups; (ii) results on the market structure variables show that high concentration and product differentiation at the beginning of the period lead to high firm entry (and exit rates or high firm turbulence); (iii) the industries where market concentration is high also have high profits that can attract new firms; (iv) in industries with strong product differentiation there will be higher firm entry, since it is easier for firms to present new products; (iv) capital intensity constitutes a barrier to entry (and exit), as does R&D intensity, even though both of these barriers are relatively low; and (v) an increase in deregulation leads on average to an increase in both firm entry and exit, but this varies from one industry to another.

Foreign direct investment (FDI) is an alternative route by which new products may be introduced and produced in an economy. There are extensive literatures exploring the main determinants of such investment flows (see, for example, Moran, 2012). The main host-country factors usually argued to attract FDI are market size, access to overseas markets, tax regimes (especially the potential to export profits) and other government policies, skill availability, university quality, capital markets, infrastructure, and regulations. The relative importance of these factors will differ by country. Blonigen and Piger (2014) use Bayesian statistical techniques to select, from a similarly large set of candidates, the variables most likely to be determinants of FDI activity. They find that variables with consistently high inclusion probabilities are traditional gravity variables, cultural distance factors, parent-country per capita GDP, relative labour endowments, and regional trade agreements. Variables with little support for inclusion are multilateral trade openness, host-country business costs, host-country infrastructure (including credit markets), and host-country institutions. Many such factors may thus be argued to impact on product innovation via FDI.

The final route for product innovations to enter a domestic market is via importation from overseas. We do not know of any literature that expressly addresses the determinants of the order in which overseas suppliers service different markets with new products. However, Comin and Mestieri (2014) provide some useful insights on a closely related issue. They use an historical data set on the diffusion of a number of technologies to explore the extensive margin of technology adoption, which indicates how long it takes a country to introduce a new technology for the first time (as well as the intensive margin that measures the spread of ownership after first adoption). Their data set encompasses 166 countries, spans the period from 1820 through to 2003, and covers major technologies such as transportation technologies (consisting of steam and motor ships, passenger and freight railways, cars, trucks, and passenger and freight aircraft), telecommunication technologies (telegraphs, telephones, and cell phones), information technology (PCs and internet use), medical technology (MRI scanners), steel produced using blast oxygen furnaces, and electricity. Most of these technologies are embodied in either consumer or producer product innovations, and thus by exploring the lags across countries in first adoption we will get some insight into the determinants of the order in which overseas suppliers service different markets with new products.

Comin and Mestieri (2014) report that the average adoption lag, across all technologies and countries, is forty-four years, but there is significant variation in the average adoption lags across technologies, ranging, for example, from seven years for the internet to 121 years for steam and motor ships. There is also considerable cross-country variation in adoption lags for any given technology, the range for the cross-country standard deviations going from three years for PCs to fifty-three years for steam and motor ships. The authors also find that adoption lags have converged across countries over time and, in particular, the rate of decline in adoption lags is almost 40 per cent higher in non-western than in western countries. This convergence means that more recent technologies have, on average, experienced faster diffusion via the extensive margin. The factors that drive the patterns in the extensive margins (and changes therein over time) are, however, not clear. Suggestions are that differences across countries in knowledge bases, trade openness, demand, human capital, adoption history, and geographic interactions have major roles to play.

Considering new start-ups, FDI, and imports as a whole, these findings from the literature would appear to suggest that there are few differences between what induces domestic product innovation and imported product innovation.

8.6 Conclusions

The initially declared aim of this chapter was to explore the empirical literature on the determinants of the extent of product innovations in firms. There is only a limited literature on this issue, much of the literature being more concerned with R&D, patenting, or process innovations. It was shown, however, by using survey data, that firms that are innovative in one dimension are also innovative in other dimensions. Thus R&D-intensive firms are also design-intensive, and firms active in process innovation are also active in product innovation. It was thus argued that (i) the main issue should be what distinguishes the innovating firm from others; and (ii) literature based on the analysis of other innovation indicators should also give some insight into the determinants of product innovation.

Relying largely upon existing surveys, the existing literature leads us to conclude that (i) technological characteristics of industries matter; (ii) firm and market characteristics are not linearly related to innovation, but rather there are many and various feedback loops between them; (iii) the two most important firm characteristics affecting innovativeness are internal finance, most notably cash flow, and sales, both being positively correlated with innovative activity; (iv) an inverted U-shaped curve captures the relationship between competition or rivalry and R&D; (v) competition in foreign markets, via exporting, is predominantly found to have positive effects on R&D and innovation at home; (vi) there are various spillover effects among firms, between universities and firms, and across national boundaries; and (vii) R&D tax credits and direct subsidies have positive effects on private R&D investment, although there is some indication that the effect of subsidies may be non-linear and also show an inverted U-shaped relationship.

These findings based upon the existing literature were supplemented by an analysis of survey-sourced data on the characteristics of the innovating firm. Here it was found, largely in confirmation of previous findings in the literature, that in the introduction of innovations (i) the industrial sector in which the firm is located matters; (i) the extent of firm innovativeness seems to increase with firm size; (iii) formalized R&D exerts a positive impact upon innovation activity; (iii) there is an important link between innovation and in-house skills; (iv) public support increases innovative activity; (v) the proportion of firms that are part of a group (vs independent establishments) is higher in more innovation-active firms than in others. In the sourcing of innovations it was found that a firm is more likely to actively source innovation if it is larger, is foreign-owned, offers both physical products and services, has a higher proportion of employees with a degree between 11 per cent and 20 per cent, has a higher percentage of turnover from exports, and has some product uniqueness. Although a warning about causality is deserved, it is clear that the empirical evidence for the characteristics of the innovative firm supports the findings that use more traditional indicators.

Survey data on the constraints to innovation were also analysed. Although a high degree of heterogeneity was found in the responses across countries, the issues considered to be of main importance (at least for the United Kingdom and some other EU countries) are (i) innovation costs; (ii) risk and finance; and (iii) the availability of qualified labour. These have also been discussed in chapter 7. Product market issues (competition under various guises) and regulations are also raised as potential barriers. The former has been covered in the discussion of exiting literature, whereas consideration of the latter will be resumed in chapter 13, where policy is discussed.

Finally, three sources of product innovations that tend to fall outside the remit of survey-based innovation studies—entry/start-ups, FDI, and imports of new products—were discussed in order to provide a fuller picture of the determinants of product innovation activity. At a high level of generality, the same factors that have been shown to affect the generation of product innovations in the domestic economy also affect product innovation by these three other sources.

9 The Diffusion of Product Innovations

9.1 Introduction

In chapters 4 and 8 we have explored the extent and patterns of the generation and launch of product innovation in some detail. The aim of this chapter is to consider empirical evidence describing patterns in the post-launch development of ownership or sales (or both) of the newly introduced products. This we label 'diffusion of product innovations'. There are several literatures that relate to the analysis of the diffusion of innovation in various disciplines. The extensive literature in economics is reviewed in Stoneman (2002) and in Stoneman and Battisti (2010). In the communications field the work of Rogers (2003) has been a major influence. Reviews of the literature in marketing, starting from Bass (1969), can be found in Mahajan et al. (2000) and Peres et al. (2009), and there is a critical review by Chandrasekaran and Tellis (2007). There is a spatial-based literature in geography (see, for example, Clark, 1984; Baptista, 2001; and Feldman and Kogler, 2010). The analysis of diffusion also has many overlaps with the analysis of the product life cycle (see chapter 4), but we do not explicitly pursue that here.

Our purpose here is not to provide another wide-ranging review of these literatures. In particular the intent is to emphasize the what rather than the why, that is, to reveal patterns rather than to concentrate upon explaining the determinants of those patterns. The prime reason for this is that, although much of the literature in all the different disciplines tends to concentrate on considering diffusion as a demand-based phenomenon, the outcomes that one actually observes, in terms of patterns of sales and ownership over time, are the results of a complex interaction between (i) demand patterns and their development over time; (ii) supply decisions on capacity, pricing, location, and promotion and changes therein over time; and (iii) further innovations in both production and product characteristics in differing horizontal and vertical directions. It is beyond the limits of this chapter to unravel these different factors in all the cases considered.

In chapter 2 we have attempted to distinguish (although with limited precision) three different types of innovations: original, new-to-market, and new-to-firm (but not new-to-market) innovations. In this chapter we will concentrate upon analysing the ownership and sales of the first two of these. Although the extent of new-to-firm innovation may be a factor that, over time, helps to extend the diffusion of technologies that are original or new to the market (for example, the launch of a generic pharmaceutical product that is cheaper may well extend the use of that remedial treatment), the actual intertemporal pattern of the sales of a specific new-to-firm (but not new-to-market) product may be difficult to observe, and in any case is of limited interest (in relation to original and new-to-market products) in the current context.

Original products, by definition, are unique when launched. As time proceeds, however, we expect further innovations of a new-to-market form to follow (although one cannot rule out new-to-firm innovations that are not new to the market). The new-to-market products may offer improvements or differences in performance or lower prices, increasing demand for the new product. As more new-to-market products appear, the extent of competition on the market will increase, and this competition itself may lead to lower prices and greater demand for the generic product. The increasing number of suppliers will also tend to generate greater capacity to supply the market (over and above the capacity that the original supplier generates), again stimulating, through lower prices, both sales and or ownership. Although, as we have seen in chapter 4, a multiplicity of new-to-market offerings may imply that a wide mix of designs and standards are available on the market at early stages in the development of a product, this multiplicity may actually confuse and deter potential buyers, hence preventing widespread use and ownership until there has been some centring upon a dominant design or standard that removes much of the market confusion and uncertainty.

The pattern whereby an original new product offering is followed by further new-to-market offerings may be illustrated by how new products often go through different generations as they mature. Thus for example the mobile phone has passed through several generations in its life (see Liikanen et al. 2004 and Bohlin et al. 2010), from analogue to digital, and then through various generation of iPhone, each being a vertical-quality improvement on the previous generation. This can be seen also in the history of computers, where mainframes went through a first, a second, a third, and a fourth generation before the advent of personal computers and laptops. The study of diffusion may either concern the overall product class (mobile phones or computers) or treat each generation as a separate area of study—there is no clear rule as to what route the student should follow, although one should note that, if one is to study the overall product class, then the generational improvements will be a driving force, whereas, if one is to study the generations separately, then the advent of the next generation may well explain the demise of a previous generation.

In chapter 2, in addition to identifying original, new-to-market, and new-to-firm product innovations, we also identified a number of different types of products. In particular, we distinguished between producer and consumer goods that are differentiated by whether the buyers are firms or households and between durables and non-durables, according to whether the product provides a flow of services over time or only at a point in time and has to be repurchased (this category includes services). Most of the diffusion literature concentrates upon durable products. Economics also concentrates upon producer durables, although marketing, for example, places more emphasis upon consumer goods. However, there is a growing literature on the diffusion of particular management practices (see Bloom and Van Reenen, 2010).

When original and new product innovations are made, they will not necessarily appear in all economies at the same time (in some economies they will not even appear at all). Economic activity is becoming increasingly globally integrated and the analysis of diffusion phenomena needs to address this. Thus, at any moment in time, the new product may be imported in some economies while in other economies it may be produced at home or even exported. One could thus think of two sides to the diffusion of a new product in an economy: (i) the initiation of ownership and the subsequent growth of demand or ownership; and (ii) the initiation of domestic production and the subsequent growth (or, as we shall see, possible eventual decline) in domestic production of the new product. To illustrate, one needs to explore international diffusion patterns. Thus the analysis here encompasses not only diffusion within single countries but also differences across countries. There is also the potential for interactions between diffusion patterns in different countries.

As a final introductory remark, we note that each new product eventually becomes an old product and will no longer be produced, demanded, or owned (after more or less time). It is in the tradition of diffusion analysis to ignore the final stages of product life and to concentrate upon the growth and maturity phase. This will also be the case here. There will also be many new products that do not succeed, as we have already stated in a number of places. In similar vein, because we concentrate on the growth phase for new products, we ignore products that have no market impact, that is, products that fail.

In the sections that follow we will continue to distinguish between production and consumption, durables and non-durables, and households and firms. We start by considering the international picture and emphasizing diffusion across countries. We then consider within-country diffusion, emphasizing differences in diffusion across and within firms or households.

9.2 The global perspective: the extensive margin

In a series of papers, for example Comin and Hobijn (2004, 2009, 2010) and Comin et al. (2006, 2008a, 2008b), historical data regarding the international diffusion of a number of largely original technologies have been presented and discussed. These data are available online in the Historical Cross Country Technology Adoption (HCCTAD) data set and in the Cross Country Historical Adoption of Technology (CHAT) data set. The HCCTAD covers nearly two and a half centuries, from 1788 and to 2001, but is limited to the leading twenty-three industrialized countries in the world. CHAT is an unbalanced panel data set with information on the adoption of over 100 technologies in more than 150 countries since 1800. The data are available for download at http://www.nber.org/data/chat. The technologies that are covered relate primarily to durable goods but include both consumer and producer durables, for example mobile phones, TV sets, different steel production technologies, steamships, and motor ships. Comin and Mestieri (2013) consider two main dimensions of the global diffusion of technology. The first tells us whether a specific technology is present or not in a given country at a moment in time. The time taken from invention of the technology to its first appearance in a country can be labelled the 'adoption lag'. The study of the differences in this lag across countries and technologies may be seen as an analysis of the extensive margin of technology diffusion. The second dimension, the intensive margin, concerns how intensively a technology is used once it is present in a country. We consider the former here and the latter in the next section.

With respect to the extensive margin, Comin and Hobijn (2010), using data on the diffusion of fifteen technologies in 166 countries over the past two centuries, find, first, that adoption lags are large. The average adoption lag is forty-seven years. There is, however, substantial variation in this lag, both across countries and across technologies. The standard deviation in adoption lags is thirty-nine years. An analysis of variance yields that 54 per cent of the variance in adoption lags is explained by variation across technologies, 18 per cent by cross-country variation, and 11 per cent by the covariance between the two. The remaining 17 per cent is unexplained. It is also found that newer technologies have been adopted faster than older ones. An acceleration in technology adoption took place during the two whole centuries covered by the data and therefore preceded the digital revolution or the postwar globalization process, which have often been cited as the driving forces behind the rapid diffusion of technologies in recent decades. It is also found that the development of Japan in the second half of the nineteenth century and first half of the twentieth century—and of the so-called four East Asian tigers in the second half of the twentieth century—all coincided with their experiencing a catch-up with the range of technologies used by industrialized countries and involved a substantial reduction in adoption lags in these countries by comparison to (other) OECD countries.

The technologies that are studied in Comin and Hobijn (2010), which include passenger transport by cars, freight transport by truck, telephones, cell phones, and magnetic resonance imaging (MRI), tend to be technologies in which production and consumption or use do not coincide. Thus, although with some technologies, such as services or certain medical procedures, the act of supply is also the act of production, in most cases innovations will involve the production and the use of a good as separate activities. Thus the use of motor vehicles (cars or trucks) differs from their manufacture. Similarly, the use of cell phones or telephones is not the same as the manufacture of the systems and handsets that enable such use. Thus for many products there are two diffusion processes that happen simultaneously, each with its own extensive margin. In the vehicle case, there is the adoption lag in the adoption of travel by motor vehicles (as studied by Comin and Hobijn, 2010); and there is also the adoption lag in the manufacture of motor vehicles (not studied). The link between the two is of course international trade. To give a historical example, for many years the United Kingdom was the only producer of stationary steam engines and locomotives, but manufacture using steam, and steam rail travel, diffused much more widely around the world. More recently, the diffusion of cell phone usage was rapid in most countries. They are, however, still manufactured in only a few. Despite these arguments, if the production of innovative goods were also considered, then the finding that adoption lags are large with substantial variation both across countries and across technologies would be reinforced rather than countered.

9.3 The intensive margin

Using data from the HCCTAD data set, Pulkki-Brännström and Stoneman (2013) have explored the relative importance to the worldwide spread of technologies of spread across countries (the extensive margin) and spread within countries (the intensive margin). They explicitly consider the case of mail services for the period from 1830 until 1990 in fifteen OECD countries. They argue that, if one defines the overall worldwide level of use of a new technology as D(t), the number of using countries as z(t), and the average intensity of use in each country, D(t)/z(t), as equal to w(t), then $D(t) = z(t) * (D(t)/z(t)) = z(t) * w(t)$—and, in growth terms, $\Delta \ln D(t) = \Delta \ln z(t) + \Delta \ln wt(t)$. This illustrates that the growth of the worldwide level of use of the new technology is made up of two parts: the growth in the number of using countries and the growth in the average intensity of use in each country. These may be measured for each period, and the relative importance of changes in the extensive and intensive margin is measured too.

Measuring diffusion by units of mail handled and defining a country as having adopted the technology of postal services when the units of mail handled per year exceeded 10 million, the authors show that by 1890 all sample countries had introduced a mail service. Thus, after 1890, there were no further advances in the extensive margin and changes therein no longer contributed to sample-wide (i.e. across countries) diffusion. Nevertheless, usage of mail services continued to grow through to the end of the studied period. This later growth was thus wholly due to extensions of the intensive margin. This is a pattern also found in other technologies. It would seem to be a stylized fact that the spread of a new technology to more and more countries (i.e. the movement of the extensive margin) is completed long before the extent of usage in each country (i.e. the movement of the intensive margin) comes to an end. Thus, although in the early stages of diffusion movements in both the extensive and the intensive margin contribute to the extensions of diffusion, in the later stages only movements in the intensive margin matter. Comin and Hobijn, (2010) also make this observation and extensively explore the intensive margin of technology diffusion.

In fact there is now a large and still growing literature that considers diffusion via this intensive margin. We should note, however, that, when looking at the use of a technology within a country, there are, once again, two margins. The first refers to the number of users (perhaps as a percentage of the population); the second refers to the intensity of use by each user. To avoid confusion, these margins (which reflect extensive and intensive margins at the international level,) are called here inter-firm (or household) and intra-firm (or household) diffusion. We first discuss the inter-firm (household) diffusion literature.

This is not the place to attempt to provide a survey of the literature on this issue. The reader is initially referred to Stoneman (2002) for the economics literature. For other disciplines, Rogers (2003), Peres et al. (2009), and Clark (1984) are informative. It is worth highlighting here a few studies that illustrate the diversity of the diffusion experiences that have been studied. There is of course work on the CHAT and HCCTAD data sets, which encompass the intra-country diffusion experiences of a wide mix of technologies and also provide international comparative data. In addition, one can mention work on the diffusion of producer durables, for example computerized numerically controlled machine tools (CNC) and computer-aided design (Stoneman and Kwon, 1994); the diffusion of consumer durables, for example mobile (or cell) phones (Liikanen et al., 2004) and home computers (Dickerson and Gentry, 1983); the diffusion of non-durable consumer and producer goods, for example unleaded petrol (Battisti and Stoneman, 2000); the diffusion of services (Barak et al., 2009); the diffusion of agricultural technologies (Griliches, 1957 and Feder and O'Mara, 1982 for developing economies); and the diffusion of organizational innovations (Antonelli, 1985 and Alange and Steiber, 2011).

Diffusion on the intensive margin often takes a considerable period of time before it is complete. On the basis of the CHAT data set, Comin and Mestieri (2010) state that their analysis of fifteen technologies and 166 countries yields the findings that there are large cross-country differences in the intensive margin of adoption and that the intensive margins vary substantially across technologies. As noted by Rosenberg (1976, p. 191), 'in the history of diffusion of many innovations, one cannot help being struck by two characteristics of the diffusion process: its apparent overall slowness on the one hand, and the wide variations in the rates of acceptance of different inventions, on the other'. However, as Comin and Hobijn (2010) state with respect to the same wide sample of technologies, their results are roughly consistent with the most remarkable finding of the traditional diffusion literature: that, for the majority of the technologies for which it has been possible to construct a diffusion measure, the diffusion curves are S-shaped. This is found in nearly all studies in nearly all disciplines. If the diffusion curve is S-shaped, if one plots the extent of usage (a measure that has to be precisely defined) against time, then, from some start date when usage is zero, usage grows initially at an increasing rate up to a point of inflection, after which growth over time, although still positive, slows down and eventually usage approaches an end point or asymptote. The precise curve is defined by the start date, the asymptote, and the implicit growth rate between the two. There is widespread evidence that these three parameters as well as the precise functional form of the diffusion curve differ both across countries and across technologies.

Much of the extant diffusion literature attempts to explain the shapes of the diffusion curves observed, potentially by addressing the determination of start date, asymptote, and slope. It is the view taken here that any explanation must encompass both demand-based and supply-based elements, which is still not particularly common in the literature. Beyond that, however, apart from one point, we resist the temptation to expand any further and refer the reader instead to Stoneman (2002), where the whole issue is examined in considerable theoretical and empirical depth.

The one outstanding point is that it is common in the diffusion literature to compare and contrast the diffusion experiences in different countries—say, the United Kingdom versus the United States, or Italy versus France. What is much less common is to consider whether diffusion in one country is affected by diffusion in others. Pulkki-Brännström and Stoneman (2013) address this issue by looking at whether a link can be found between diffusion in one country and diffusion in the rest of the world. Two possible links are suggested. The first is that experiences in other countries may yield information that can impact upon diffusion at home. Little support is found for this hypothesis. The second hypothesis is that, when final products are sold on a world market, diffusion overseas may impact upon world prices, which will change the incentives towards the further adoption of a new technology. Empirical support is found for this hypothesis.

9.4 **Intra-firm and intra-household diffusion**

Changes in the extent of use of a new technology in a country may reflect both extensions to the number of users and changes in the intensity to which the innovation is used. Although there may be some technologies where only one unit will ever be purchased by a household or firm (e.g. a particular music recording or book), in other cases there may be products for which many units are purchased. Clearly, if the product is non-durable, multiple purchases are likely; but, even with durables, households and firms may buy many units. The literature on the diffusion of non durable products, for example unleaded petrol (Battisti and Stoneman, 2000) or new services (Libai et al., 2009), does not itself add a great deal to the discussion in the previous section. However, the literature on intra-firm or intra-household diffusion of a durable product yields more that is of interest. Although this issue has been studied in economics from the earliest days—an example is Mansfield's (1963) study of the spread of diesel engines replacing steam locomotives in US railroads—in general the literature has placed much less emphasis on intra-firm (or household) diffusion than on inter-firm (or household) diffusion.

One of the more interesting issues concerns the relative importance of inter-firm and intra-firm diffusion of innovations in the overall diffusion process. Battisti and Stoneman (2003), using primarily the example of CNC in the UK metalworking and engineering industry, show that, even many years after the date of first use of a new technology, although the larger proportion of firms may be using that technology, because the extent of intra-firm diffusion is limited, the proportion of total industry output produced on that technology may be quite small. The percentage contribution of extensions of intra-firm diffusion to increases in the overall diffusion of CNC (output produced using CNC) for the periods 1975–80, 1980–5, 1985–90, and 1990–3 was increasing from 35.3 per cent to 58.9 per cent to 86.7 per cent and finally to 90.0 per cent. Backward projection techniques show that intra-firm diffusion tends to lag behind inter-firm diffusion over the whole of the diffusion process and inter-firm diffusion patterns and extent may therefore be poor indicators of overall diffusion, especially in the latter stages of the diffusion process. Understanding intra-firm diffusion is therefore crucial to the understanding of overall diffusion patterns.

Battisti et al. (2009), using data from the third UK CIS, model the usage of e-business across and within firms in the United Kingdom in the year 2000 as a single observation upon an integrated process of inter-firm and intra-firm diffusion. The data indicate, inter alia, that inter-firm usage is an incomplete indicator of the overall usage of a new technology; for, although in 2000 83.2 per cent of firms considered themselves to be e-business users, only 26 per cent are using the technology at anything other than a basic level.

This justifies looking at both margins. In addition, inter-firm usage is not necessarily a reliable indicator of the extent of intra-firm usage and thus data on the latter contain new information. Battisti et al. (2009) go on to determine that the pattern of e-business usage reflects the heterogeneity of firms in terms of size, other innovative activity, and labour force skills (generating differences in the payoffs to use), as well as market and non-market intermediated externalities.

Further examples of the intra-firm diffusion phenomenon in economics can be found in Battisti and Stoneman (2005), looking at CNC technology, Battisti et al. (2007), looking at IT, and Battisti and Iona (2009), looking at management practices, while Steffens (2003) is an example in marketing. Further discussion of the possible drivers of intra-firm diffusion can be found, for example, in Romeo (1975), and Stoneman (1981).

9.5 The diffusion of the production of product innovations

Despite the breadth of the empirical analysis of diffusion phenomena, it is still very much the case that, just as with the extensive margin, the diffusion literature dealing with the intensive margin has concentrated upon the use or adoption of particular technologies as opposed to the production of the products that embody those technologies. Thus, for example, there are studies of the diffusion of both black-and-white (B&W) and colour TV ownership (e.g. Bain, 1962; Stoneman and Karshenas, 1992), but as far as we are aware the changing pattern of TV production across and within countries has not been as much studied in the diffusion literature. Such issues have been addressed, however, in a particular line in the economics literature known as 'North–South models' (see Afonso, 2011 for a useful review), which has emphasized that the developed economies (the North) account for most of the worldwide innovations (see for example Eaton and Kortum, 1999, 2001), but that over time technological knowledge is diffused to other developing economies (the South), thereby spreading the benefits of innovation worldwide. As this spreading occurs, the manufacture of innovations should be moving from the developed to the developing countries. We made some comment on this at the end of section 9.2, but here we present some data on the diffusion on the intensive margin of the production of goods that embody new technologies.

Television is a good example. Gao and Tisdell (2004) have explored the changing patterns of the worldwide production of both B&W and colour TV sets. They find that, during the mid-1940 to late-1950 period of B&W, the United States and the United Kingdom were the major producers of

TV sets (producing 91 per cent and 8 per cent of sets respectively in 1948 and 76 per cent and 13 per cent in 1953). The two were joined from the mid-1950s on by other European countries, and then by Japan. Thereafter the US and UK shares declined significantly to around 28 per cent (United States) and 11 per cent (United Kingdom) in the early 1960s. In colour, the United States was the innovator; Japan and EU countries followed. Early imitators such as France, West Germany, Italy, and Japan all experienced an early increase and then a decline in production. In the late 1960s, Japan took the lead from the United States; its share rose dramatically in the 1960s until the mid-1970s and it became the largest colour producer, a position it held for almost two decades; but its share had fallen to only 7 per cent by 1995. US and UK shares fell to 14 per cent and 3 per cent respectively in the 1980s. During the period from 1960 to 1970, production commenced in Taiwan, South Korea, Mexico, Italy, and Malaysia, all relatively low-income countries. However, from the early 1980s onwards, China dramatically increased its TV production and in 1987 it became the largest single producer globally.

We make no attempt to explain these patterns. They might, for example, reflect the behaviour of multinational companies, government industrial strategies, differences in labour costs, the movement of skilled labour, or many other factors. The point to make is that, as the diffusion of the use and ownership of TV sets was taking one path, the manufacture of TV sets was taking quite another path. In fact, as manufacturing shifted from the United States and the United Kingdom and their TV set production industries went into decline, the diffusion of TV production was over for these countries. From then on diffusion in use depended largely on imports.

Two further examples illustrate the same point. In 1967 the worldwide manufacture of steel was 497.2 million metric tons, of which 23 per cent were produced in the United States, with almost negligible production in China (all data are taken from https://en.wikipedia.org/wiki/List_of_countries_by_steel_production). By 2000 world production was 850 million metric tons, of which 12 per cent were made in the United States, 23 per cent in Europe, and 15 per cent in China. US production had fallen by 14 million metric tons since 1960. In 2015 world production was 1,620.4 million metric tons. The US share was 5 per cent and its production had fallen by 22 million metric tons since 2000; the China share was 50 per cent and its production had increased by 675 million metric tons since 2000. The EU share had fallen to 10.2 per cent (a fall in production of about 30 million metric tons), and the South Korean share was 4 per cent.

In 1960 the worldwide production of passenger cars, light commercial vehicles, minibuses, trucks, buses, and coaches was 16.5 million units, of which 37 per cent were made in the United States, 12 per cent in Germany, 11 per cent in the United Kingdom, 3 per cent in Japan, and a negligible number in China (all data come from https://en.wikipedia.org/wiki/List_of_countries_by_motor_vehicle_production). By 2000 worldwide production was 58.4 million units, of which 22 per cent were made in the United States (although production was still above the 1960 level), 17 per cent in Japan, 9.4 per cent in Germany, 3 per cent in the United Kingdom, and 3.5 per cent in China. In 2015 worldwide production was 90.7 million units, China producing 27 per cent, the United States 13.3 per cent (a fall of almost a million units since 2000), Japan 10 per cent, Germany 6.6 per cent, South Korea 5 per cent, and the United Kingdom 1.8 per cent.

Jointly these examples illustrate that the diffusion of the production of new products may well involve shifts in production from one country to another over time, both relatively and in some cases absolutely. As already stated, it is not our intention here to explore causes. However, there are obvious clear parallels between such changes on the international scene and the discussion we have already had in section 4.5 on the issue of industry maturity and product innovation.

9.6 **Conclusions**

In this chapter we have explored the patterns of adoption and use of original and new-to-market product innovations with the intent primarily of showing how the use and ownership of these innovations and the production of the goods that embody such innovations develop or diffuse over time. We have identified three levels of diffusion: the spreading of first use across countries (the extensive margin); the spreading of first use across users within countries (the intensive margin or inter-firm/household diffusion); and the increasing intensity of use by adopters (intra-firm/household diffusion). The details encompass both producer and consumer goods, durable and non-durable products, and even management practices. Although the emphasis is upon the economics literature, there is also reference to the literatures in marketing, management, and geography.

The principal finding of the literature is that diffusion takes time, often a considerable period of time. The average adoption lag on the extensive margin is of forty-seven years, although there are considerable differences across technologies. Similarly, the spread of product innovations within countries is time-intensive and differs considerably by technology and country. Such movement on the intensive margin often continues for many years after diffusion on the extensive margin is completed. Intra-firm or household diffusion is also time-intensive, differs by industry sector, country, and technology, and continues even after inter-firm or household diffusion is complete. We have also observed that the diffusion of the production of product innovations may eventually mean that countries that were early producers are eventually replaced by producers in countries that were late in the adoption process. The diffusion of the new may thus be reflected in the withdrawal from production of the old.

Finally, the main purpose was to emphasize the what rather than the why—that is, to reveal patterns rather than to concentrate on explaining the determinants of those patterns. We consider, however, that, although much of the literature in all the different disciplines tends to treat diffusion as a demand-based phenomenon, the outcomes that one actually observes in terms of patterns of sales and ownership over time are the result of a complex interaction between (i) demand patterns and their development over time; (ii) supply decisions regarding capacity, pricing, and promotion and changes therein, over time; and (iii) further innovations in both production and product characteristics, in differing horizontal and vertical directions. We have not tried here to unravel these different factors in the cases considered.

10 Product Innovation and Firm Performance

10.1 Introduction

The purpose of this chapter is to explore the relationship between the product innovation activity of a firm and its performance. Although some alternative performance indicators such as productivity and exporting behaviour will also be considered, the chapter will primarily concentrate upon measures of firm performance that are related, either closely or loosely, to profit indicators. This emphasis reflects a balance in the literature that is not surprising. Profit is not only the sine qua non of firm survival but it is also, to a greater or smaller degree, the main driving force behind entrepreneurial activity in market economies.

Product innovation will impact upon profit in a number of ways: the costs of developing and introducing a new product will reduce the innovator's profits; increased revenues earned by selling more with a new product, or by selling such a product at a higher price than the old, will lead to higher profits; the costs of producing the new product (i.e. the productivity of the firm) may be lower than those of producing the old one, which implies higher profits, or higher, which implies lower profits; and product innovation by others may have a negative impact upon a non-innovator's profits. In theory, although alternative scenarios are sometime considered, it is generally argued that successful product innovation by a firm will have a positive effect on the innovator's economic performance (although the argument may be circular in that 'success' is often defined by the extent of profit gain).

However, profit gains may be temporary. Thus for example, in the Schumpeterian view of competition (Schumpeter, 1934), product innovation may enable firms to gain higher profits in the short run, but these profits attract other innovators and, as the number of competitors increases over time, profits tend to diminish, such that, it is claimed, the profit gains from product innovation do not persist in to the long run. It is argued, however, that, although returns to single innovations may be transitory due to such competitive processes, firms may persistently maintain high profitability levels by introducing multiple innovations. Previous economic research in the field of innovation suggests that continuously innovative firms may maintain persistent profit differences by comparison not only to non-innovators but also to those firms that innovate only occasionally (Geroski et al., 1993;

Roberts, 1999; Cefis and Ciccarelli, 2005). We also reflect recent advances in the literature that underline how product innovation should be considered in the context of a firm's overall innovativeness, in that complementarities between technological and non-technological innovations should be considered in the determination of performance, an approach that we already highlighted in chapter 8 by introducing the concept of the innovative firm.

The importance placed on the introduction of multiple product innovations (both over time and at a point in time) has shifted some of the focus in the literature from the firm level to the intra-firm level of analysis. Studies appearing in the organization-oriented tradition have analysed extensively the black box of the product development process, placing an emphasis on the organizational characteristics, roles, and processes that determine successful innovation in firms (Brown and Eisenhardt, 1995; Krishnan and Ulrich, 2001, and Hauser et al., 2006). The growth in this intra firm literature has led us to the conclusion that in this chapter we should consider both economic and managerial perspectives[1] if we are to provide a thorough description of the mechanism underlying the link between product innovation and firm performance, although, given our main purpose, we concentrate upon the economic approach.

Exemplifying an innovation management perspective, the evaluation of product success through appropriate performance measures crucially depends on decisions about how the company will be innovative and how it will generate value from innovation (see Davila, Epstein, and Shelton, 2012). Various suggestions for constructing appropriate measurement systems are available from quite different analytical frameworks (Adams et al., 2006). One useful approach employs a model that synthesizes different dimensions of the innovation process in a causal relationship. The traditional inputs–process–outputs–outcomes model (Brown and Svenson, 1998) describes the innovation process as a system of functions where inputs (people, ideas, equipment, funds, information) are combined in processes (projects and portfolio management) with outputs (patents, products, processes) that lead to outcomes that can be expressed in both financial and non-financial metrics. In this framework, at different stages, firms manage a variety of activities ranging from single-innovation projects to a whole innovation portfolio, that is, all the innovation projects that are in the firm's pipeline, the final stage, outcomes, being a company-level indicator by which firm's performance should be assessed.

Despite the apparent completeness of such an approach, it does not integrate well with the microeconomic analysis of innovation. The metrics traditionally used in economics-oriented studies do not effectively capture the relevant

[1] A simple but basic difference between the two approaches is that economic investigations typically use empirical investigation based on econometric techniques, whereas managerial studies more often—although not exclusively—use a case study approach.

aspects of the framework, such as differences in product launch timings, expected market impacts. risk levels of ongoing projects, and quality and usage of newly launched products. The metrics primarily studied in economics-oriented studies are more concerned with capturing financial outcomes and incorporate such variables as profits, sales growth, capital values, revenues, or the share of sales generated by new products. However, given the primarily economic approach adopted in this volume, our main emphasis is upon this economic perspective rather than the managerial ones. In particular, we link the two approaches when analysing the product life cycle (PLC) and product portfolio issues, both aspects of a firm's competitive strategy that are crucial to understanding its success.

In the sections that follow we first discuss alternative measures of the impact of innovation on firm performance, then we move on to look at empirical results related to such impacts, encompassing profit indicators, market value, productivity, and export behaviour. This is followed by a discussion of complementarities before more managerial topics—such as product portfolios and life cycles, product differentiation and market orientation—are introduced.

10.2 Measuring firm performance

Measuring the outcomes of innovation is not a simple or easy task. For example, at the firm level, at any one time, there are typically different innovation projects at different stages of development underway, and a good measurement system should take into consideration all such innovative efforts in any calculation of a measure of performance at the firm level. Firm-level financial indicators are probably the most widely used. These tend to aggregate across different innovations that are developed, introduced, or exploited at any point in time. The two most widely employed financial indicators are accounting-based measures and market-based measures. The difference between them lays in the fact that accounting-based measures, which are derived from accounting information, reflect a firm's past performance, while market-based indicators may reflect anticipated future performance. There are, however, other financial measures that can be used as well, such as cost-accounting measures. Non-financial measures can and have been used too.

10.2.1 *Accounting-based measures*

A common feature of most of the economic analysis of the performance–innovation relationship is the extensive use, directly or indirectly, of accounting measures of a firm's profitability. Among the most widely used accounting measures are the return on assets (ROA), measured as after-tax operating income as a percentage of total assets; the return on capital (ROC), measured as after-tax operating income as a percentage of long-term capital; and the return on equity (ROE), measured as net income as a percentage of shareholders' equity. However, the return on sales (ROS), measured as net income (before interest and tax) as a percentage of sales, may be preferred because, as a measure of operating profitability, it pinpoints the relationship between innovative activity and a firm's core business. The other measures of firm profitability (ROA, ROE) are strongly influenced by financial returns that, first, are not directly related to the core business and, secondly, may crucially depend on the economic valuation of total assets, which in many cases may be biased. Economic value added (EVA), measured as after-tax operating income minus the cost of capital employed (and also called residual income), is considered by some a better measure of a company's performance than is accounting income, as it takes into account all the costs, including that of the financial resources invested. Compared to other accounting ratios, it is an absolute measure, thus allowing one to consider individual projects as well as the aggregate firm level. The general interpretation is that, if EVA is positive, then the firm's current operations are adding value for shareholders.

Among the different measures available, accounting ratios and the EVA approach have the advantage of showing current performance and, thus, of not being affected by financial market fluctuations. Another advantage is that they can be computed not only for the entire firm but even at the plant or division level. However, as often pointed out in the considerable literature that discusses and uses accounting profits as an appropriate measure of a firm's economic performance (see, for example, Fisher and McGowan, 1983; Benston, 1985; Edwards et al., 1987; Lindenberg and Ross, 1981; Smirlock, Gilligan, and Marshall, 1984; Connolly and Schwartz, 1985), accounting profits may be a noisy measure of a firm's economic performance.

One possible bias relating to accounting profits is that relevant ratios may depend on aggregate book values for assets, debt, and equity, while there could be specific assets that are important to the firm's growth but their value is not shown on the balance sheet. This is the case for non-financial, non-physical assets such as brands, product designs, and services or other intangible assets that are the result of investments in knowledge creation (organization or business process improvements). If computed and disclosed, the book values of assets, capital, and equity would increase. An additional drawback of accounting-based ratios is that they are not necessarily based on the current market values of the firm's assets. The assets in a company's books may be valued at their original cost, minus any depreciation—and thus older assets may be undervalued relative to current market conditions.

10.2.2 *Market-based measures*

There are three main market-based measures that have been employed:

- Tobin's Q, calculated by the ratio of the market value of the firm's shares (market capitalization) plus total debt divided by total assets of the company, where a firm's market capitalization is equal to share price times number of shares outstanding;
- market value added, measured as the difference between the market value of the firm's shares and the amount of money that shareholders have invested in the firm; and
- the market-to-book value of the firm, measured as the ratio of market value of equity to book value of equity—a measure of how much value has been added for each euro that shareholders have invested.

However, although the market-based measures will reflect forward-looking expectations of a firm's performance, in the case of complex innovative investments it could take many years for any economic returns to be produced. If they are based on the current market value of a firm's assets, market-based measures may be preferable to accounting-based measures, but only under the implicit and strong assumption of efficient financial markets. However, market values may fluctuate because of risks and events that are outside the financial manager's control, and thus even market-based measures may be biased and fail to describe how well the firm's management is actually performing. Another consideration is that the market value of smaller, privately owned companies is not observable, as their shares are not traded. The same is the case for the market value of single plants or divisions that are parts of larger corporations. In all these cases, accounting measures of profitability are the only possible solution to financial performance measurement.

10.2.3 *Cost accounting measures*

Kuczmarski (2000) and Kandybin (2009) propose the return on innovation investment (ROI2) as a metric that provides a reliable analytical tool that can aid in the management decision process and ultimately discriminate between successful and unsuccessful innovations. In particular, Kuczmarski (2000) defines ROI2 as the ratio of a firm's total profits from new products divided by total expenditures on developing new products. These expenditures include research costs—that is, market research, customer research, and concept testing; development costs—that is, product definition, design, prototype development, and actual product development; and initial commercialization and prelaunch costs, which in turn cover market testing, commercialization development, and media costs. It is argued that the firm should focus on the success rate of its total product portfolio so that possible product failures may be compensated for by other product success. The new product survival rate, that is, the ratio between the number of launched products still on the market and the total number of products launched over a specified period, is an alternative-related measure that would indicate whether the firm's products lasted long enough to recover costs and thus earn profits. Kandybin (2009) expanded the ROI2 concept by introducing the so-called innovation effectiveness curve, which plots annual spending on innovation against the financial returns from innovation projects (with returns expressed in terms of a project internal rate of return). Kandybin (2009) considers, on the basis of empirical investigation, that such a curve appears to be stable over time and is positively correlated with a firm's growth. There are three distinctive sections of the effectiveness curve: the hits, the healthy innovation, and the tail. The first section involves a relatively small number of projects (firms) showing very high returns; the second, which incorporates the bulk of projects, refers to innovation projects that provide economically significant returns; the third shows the low-return or no-return projects, and therefore firms that can be defined as strugglers. It is worth noting that the curve may shift either upward or inward and that an increase of, say, R&D effort alone will not necessarily produce an upward shift of the curve.

The ROI2 approach to measuring a firm's innovative performance is thus a multivariate (multidimensional) approach. Kandybin (2009) identifies seven key performance indicators that should be targeted if long-term growth potential is to be properly exploited: the average internal rate of return for innovation projects (weighted by costs); total returns on innovative investment per year; total annual innovation investment; the proportion of the portfolio made up of projects with low returns (the tail); the ratio of projects designed to maintain market share; and the average forecasted revenue of 'high-risk, high-reward' projects (the so-called 'big idea'). It is thus suggested that firms can improve their innovative performance by looking at the components of ROI2, and, by choosing the appropriate innovation portfolio, implement appropriate strategies that can produce the desired outcomes.

10.2.4 *Other financial measures*

In addition, there is a considerable amount of literature in the economics tradition that relates firm productivity growth to innovation, where productivity can be measured as the ratio of a firm's value added to the total number of workers and/or the capital stock employed. We review this literature. An alternative indicator used in economics relates to the exporting behaviour of the firm. It is argued that successful innovators are successful exporters. We also consider this further. Other indicators that have been used in the literature and that we address *en passant* relate to (i) growth in the firm's sales that results from innovation; and (ii) the firm's ability to translate innovation input into innovation output, that is, the productivity or technical efficiency of the innovative process.

10.2.5 *Non-financial (subjective) measures*

The use of subjective measures of economic performance is widespread within the managerially oriented literature. These measures, which mainly reflect managerial perceptions, are typically based on appropriate surveys where respondents are asked to assess their respective company's performance vis-à-vis that of their competitors (Harris, 2001). In an influential paper, Venkatraman (1989) suggests sales' growth position, satisfaction with sales' growth rate, and market share gains as operational indicators of a firm's growth. In addition, he proposes satisfaction with return on corporate investment, net profit position, return on investment (ROI), satisfaction with return on sales, and financial liquidity position as individual indicators of a firm's profitability.

Although non-financial methods of economic performance evaluation are appropriate when financial information is missing or incomplete or when it is crucial to provide timely information, the major problem with these methods is that they may be highly subjective and hence produce biased results. In this context, it is worth noting that the overuse of subjective performance measures in the organizational studies literature is suggested by a number of authors to be a drawback (e.g., Greenley, 1995; Chang and Chen, 1998; Slater and Narver, 1994; Kerssens-van Drongelen and Cook, 1997; Harris, 2001).

10.3 **Some estimates of the impact of product innovation on firm performance**

10.3.1 *Product innovation and profitability*

A number of empirical studies have proved the existence of a positive relationship between the introduction of innovative products and firm performance as measured in terms of profitability. For example, Roberts (1999), using a Schumpeterian framework, emphasizes the role of product innovation in determining a firm's abnormal profitability. The hypothesis is that a firm's persistent (above-normal) profitability may be explained, on the one hand, by its ability to introduce new products and, on the other, by its ability to fight off competition by introducing successful multiple innovations over time. The latter explanation depends on the assumption that imitation by others tends to reduce the innovators gains from new products. In his contribution, Roberts

(1999) tests a modified version of the profit autoregressive model of profitability proposed by Geroski (1990), which allows one to estimate both the long-run level of firm profitability and the speed of adjustment towards the long-run level. Applying this model to the US pharmaceutical industries, he estimates the effects of innovative propensity on profit (i.e. the ROA) dynamics.

The originality of this approach is that it employs measures of a firm's innovative propensity on the basis of the examination of its product portfolio. Using a panel of 42 US pharmaceutical firms (Compustat and GlobalScope databases) over the period 1977–93 together with a product-level database supplied by Intercontinental Medical Statistics America, Roberts (1999) identifies a total of 1,070 drug products new to the market during the period under study. Among these new products, only 145 are defined as innovative, the selection being determined by the examination of their initial market share. The new products with relatively higher market shares at the time of their introduction are considered innovative. Following this argument, the firm's innovative propensity is defined as the proportion of sales derived from the firm's innovative products averaged across the sample period.

The results provide support for the hypothesis that a higher innovative propensity significantly influences a firm's long-run profit levels and the rates at which its high profits converge to those levels. The estimates show that long-run profitability—ROA—increases from 0.08 to 0.40 for firms with above-average innovative propensity by comparison to firms with below-average propensity. An even higher differential is observed in a subsample of high-profit firms (from −0.22 to 0.73).

This study, although original in the extent to which different product innovations are associated with a single firm, offers only a partial view of the world, because it examines the profit–innovation relationship only in a specific sector, the pharmaceutical industry, and this is a sector where the propensity for product innovation is traditionally higher than in other manufacturing industries. The approach could usefully be extended to other manufacturing sectors.

In another contribution, Roberts (2001) stresses the need for a shift from the firm level to the sub-firm level of analysis in the study of relations between innovation activity and profitability. This change should have the same importance as had the shift from an industry-level to a firm-level study of profitability in the past. Roberts recognizes that the task is hard because there is very limited access to analytic account information, such as single-product data on revenues or costs, which may enable one to estimate product-level profitability. An alternative solution may be a matching between firm-level data on profitability and data describing product-level dynamics.

In previous work and in quite a different framework, Geroski et al. (1993) used a panel of 721 UK manufacturing firms (from the Datastream Databank) observed during the period 1972–83, of which 117 produced at least one innovation. In their model, profits, as measured by ROS, are regressed on three different measures of innovative activity together with other industry- and firm-specific variables (market share, import intensity, an index of unionization, industry concentration, the interaction between market shares, and concentration). The first of the three measures of innovative activity is firm-specific and consists of the number of innovations introduced in each year up to six; the other two, number of innovations produced and number of innovations used, are designed to capture knowledge spillovers within the two-digit industries to which the sample firm belongs.

The main results may be summarized as follows. (i) The long-run effect of product innovation on a firm's profitability is positive and insensitive to the estimation method used: in a specification that includes the effect of other firm- and industry-specific determinants—and also the effect of previous innovations produced—the estimated effect is an increase of about 16 per cent over the mean profitability, ROS. (ii) This long-run effect varies across industries, but in a way that is not correlated with the number of innovations introduced in each sector during the entire period. (iii) Permanent profitability differences between innovative and non-innovative firms, which are associated with firm-specific competitive abilities, exist and are significant:[2] these are captured by fixed effects in the model specification, whose mean estimated effect is 0.48 for innovative firms and 0.30 for non-innovators. (iv) Transitory effects are small, negative, and cyclical, suggesting that innovative firms are more able than the non-innovative ones to face periods of recession.

These results are consistent with the view that permanent profitability differences between innovative and non-innovative firms arise from the set of specific skills accumulated over time by the innovative firms. This bulk of competencies, which is continuously accrued, shift attention from a product view of innovation, whose effects are transitory because they are directly connected with the timing of specific innovations, to a process view, whose effects are permanent because they are due to the superior competencies of the innovative firms. This is consistent with the concept of the innovative firm discussed in chapter 8.

The results of Roberts (1999, 2001) and Geroski et al. (1993) specifically recognize the role of managerial abilities in determining profitability and, possibly, its persistence. This conclusion reinforces a classical argument within the industrial organization literature, that firm-specific effects as opposed to industry effects play a significant role in explaining a firm's performance (Schmalensee, 1985; Rumelt, 1991). Along with this stream of empirical literature, Hawawini et al. (2003) try to verify whether previous results in favour of the role played by firms' specific (innovation) abilities are sensitive

[2] These differences are calculated by estimating the model previously used for the whole sample separately (i) for the sample of 117 innovative firms and (ii) for the sample of 604 non innovative firms.

to the specific measure of economic performance used. They base their analysis on three different measures of profitability. Two are measures of operating profitability: one is the standard accounting measure, ROA; the other is the ratio of economic profit to capital employed (EP/CE), which measures a firm's ability to create value per dollar of capital employed. The main difference between the ROA ratio and the EP/CE ratio is in the numerator of the second ratio, which is obtained by reducing income by a charge for the cost of capital employed to produce that income. The third measure of profitability is based on the notion of the market value of the firm and is given by the ratio of the total market value to capital employed (TMV/CE). Given that the numerator is the sum of the market value of equities and debts, this ratio measures the ability of management to add value to the capital invested by shareholders and by debt holders. The relative importance of firm-specific and industry-specific effects in the determination of each measure of profitability is explored. The sample data set—derived from the Compustat data source—consists of 5,620 observations for 562 firms across 55 industries observed during the period 1987–96. The model employed is descriptive and similar to those used by Schmalensee (1985) and Rumelt (1991). A general conclusion is that firm-specific effects explain a larger proportion of the variance in profitability than total industry effects, and this result holds whichever measure of profitability is used. Firm effects vary from 27.1 per cent to 35.8 per cent of total variability, while industry effects range from 10.7 per cent to 14.3 per cent.

To assess whether these general results could be extended to each firm in the industry, the same descriptive model is applied to a subsample of 342 firms that excludes a limited number of exceptional firms (value creators) and loser firms (value destroyers) in a given industry. The analysis of the firm-specific and-industry specific effects in this new sample allows one to determine the impact of a few strategic groups on intra-industry performance. The results suggest that industry factors are now more important than firm specific factors in explaining the firm's performance: for example, in terms of ROA, industry effects explain 20.1 per cent against 16.7 per cent for firm effects. This evidence may suggest that, for firms with average managerial capabilities, industry structure matters more for profitability than it does for leader and loser firms. The direct implication is that firm-specific factors are important in explaining a firm's profitability because of the presence of a limited number of firms which deviate, in their innovative abilities, from the rest of the industry.

In a cross-sectional perspective, using the same analytical framework as the one incorporated in the Crepon, Duguet, and Mairesse (hereafter, CDM, 1998) modelling approach to the innovation-performance relationship, Loof and Heshmati (2006) provide an empirical investigation of how product innovation may affect sales margins measured by the ratio of profit after depreciation to total sales. With data related to Swedish firms it is found, somewhat contrary to most literature, that product innovation positively affects the sales margin only in the manufacturing industry, which has a low estimated elasticity of 0.020. However, it is also found that that productivity significantly increases with product innovation for both manufacturing and services firms, with estimated elasticities that are quite similar (respectively 0.121 and 0.093).

Overall, the empirical evidence provided in this section suggests that product innovation is a driver of a firm's profitability both in the short and in the long run. However, a precise estimation of the expected impact is controversial, due to (i) the heterogeneity of the methodological approaches and (ii) quite different proxies for a firm's profitability that are used in the model specifications.

10.3.2 *Product innovation and the market value of firms*

The development of new products is an investment that will often take time and also yield returns over time and at later dates. Such investments should thus be evaluated according to the present value (PV) of the net stream of revenues generated in relation to the initial capital cost. The use of financial market valuation represents a valuable tool for such exercises (Fisher and McGowan, 1983). Although it would be a very informative exercise to explore the impact of product innovation on the market value of firms, most literature of this type has concentrated mainly upon testing the impact of R&D and other measures of innovative activities (e.g. patent activities), which, although related, are not necessarily good indicators of the extent of product innovation (although patenting is more likely to reflect product innovation than would R&D). In addition, the implicit assumption most frequently made in such studies is that capital markets operate efficiently. Under such circumstances, market values reflect the discounted values of future returns from innovative activities. This hypothesis has always been controversial and is even more a matter of dispute now, after the 2008 financial crisis. Such investigations also require appropriate data sets that include firms' market values. For this reason, most of the empirical literature deals with US and UK data, as the relative number of listed companies is higher in these contexts than, say, in continental Europe.

Hall (1999) provides a comprehensive survey of the literature, underlining how publicly traded corporations may be viewed as bundles of tangible and intangible assets whose values are determined by the financial markets. Thus a firm's market value may be thought of as a function of the set of assets that constitute the whole company's book asset value. The impact of R&D investment and patents may be evaluated using such an approach. Results, however, show a wide range of impacts, with the estimated coefficient of the R&D

expenditure to total assets ratio being typically not stable over time, in either US or UK markets. Also, industry effects should be considered. In particular, the impact of R&D expenditure on market value ranges from 2.5 per cent to 8 per cent in the surveyed studies, suggesting that such estimates depend crucially on the characteristics of the sample concerning time span, industry distribution, and empirical specification. The impact of the R&D stock (measured as the net present value of the sum of the depreciating flows, usually assuming a depreciation rate of 15 per cent) varies between 0.5 per cent and 2 per cent.

Further to this, Hall and Oriani (2006) provide additional evidence that includes continental European companies in the sample, thus yielding evidence on economies other than the United States and the United Kingdom. Their estimates suggest that the market value of R&D in three European countries (France, Germany, and Italy) over the period of observation (1989–98) should take into account the effect of shareholder control. The results suggest that the impact of the R&D stock on firms' market value is negligible. This result might be justified on the grounds that the presence of a major shareholder (more than a 33 per cent share) and a well-established civil law do not enable outside investors to expect future gains from possible investment in the company.

When shareholder control is not taken into consideration, the coefficient of R&D capital is positive and significant for Germany and France in the specified market value regression. However, the R&D valuation is significantly lower than the corresponding valuation estimated for the United Kingdom. In particular, France and Germany show an R&D coefficient that is respectively 0.28 and 0.33—when a stock measure of R&D is considered—by comparison with a corresponding value for the United Kingdom of 0.88. US companies show an R&D valuation that is similar to that of the continental Europe companies, although the coefficients obtained by similar analysis for different time periods (Hall, 1993) show greater impacts.

These findings are confirmed in a UK study by Toivanen et al. (2002), who use a large panel of British companies to evaluate the impact of R&D and patent activity on their market values. The authors find that a firm's market share does not affect the market valuation of R&D investment. They also find that patent activity has negative impact. This result, which is counter-intuitive and contrary to the evidence from the United States, may be related to difficulties in exploiting technological opportunities derived from patenting. In other words, this result may be an indication of the problem of appropriating returns from innovations.

A different approach is used by several studies that aim to analyse directly the impact of new product developments on the market value of firms. These studies view new product announcements as strategic tools that may enhance a firm's competitive advantage and ultimately its market value. The study by

Chaney et al. (1991) is a significant contribution in this respect. Their analysis is based on a sample of 1,685 new product announcements made by 631 US listed companies on either the American Stock Exchange (AMEX) or the New York Stock Exchange (NYSE) for the period 1975–84. The analysis addresses the issue of whether there is a significant difference in firms' performance according to their attitudes to introducing new products. The study also aims at estimating the difference in returns to new products across industries and/or firm type (e.g. size) and across different types of innovative behaviour, for example true innovators vis-à-vis imitators.

Firms' stock price behaviour is analysed using the traditional capital asset pricing model (CAPM), from which firms' expected returns may be derived from a typical 'beta' regression that correlates a firm's stock price with the average market stock price. In other words, the hypothesis under investigation implies a test of whether information concerning new product launches affects excess returns (with respect to the market average). Results suggest that innovating firms do show better market performance than non-innovating companies. This pattern is more significant for technologically based industries, such as computers, chemicals, pharmaceuticals, and electrical equipment. In addition, the performance effect is more relevant for the introduction of original new products than for simple product changes or repositioning. However, one has to take into account that simple product updates are undertaken by only a relatively small number of firms in the sample, and therefore this evidence should be considered as a preliminary indication that needs to be supported with more evidence. Finally, firm size is found to not have a clear-cut impact on a company's excess return, although, once industry effects are controlled for, there is some evidence that firm size has positive impact.

A similar approach is used by Fehle et al. (2008), who study the impact of a brand's value on a firm's stock market value. Their analysis considers those international brands that appear to be the most valuable according to the Interbrand list, from the time of its first publication in 1994 until 2001.[3] The authors used the approach proposed by Fama and French (1993) to verify whether the difference between a firm's stock return and the free-risk return is associated with (i) the market differential (the traditional beta) and other firms' characteristics; (ii) a difference in returns between small and large firms; (iii) a difference in returns between high book-to-market ratio firms and low book-to-market ratio firms; and (iv) a momentum index.[4] They found that the sample of companies with the 100 most valuable worldwide brands (WMVBs) performed comparatively better over the period of investigation (1993–2000) than either a restricted or an enlarged sample of companies

[3] Interbrand is a private consultancy company that has published Best Global Brands since 1994.

[4] Momentum is an acceleration of a stock price and the authors adopted the measure proposed by Fama and French (1993), i.e. the so-called UpMinusDown factor (UMD).

(i.e. all listed companies at NYSE). Such a result was derived from a sort of augmented CAPM regression, in which one can evaluate the so-called beta for the most valuable brands and their average performance with respect to the risk-free return. Adjusting the analysis for industry heterogeneity, that is, decomposing the WMVB list into the brands' twenty-eight two-digit industry classification (SIC code) does not change the results, which suggests that the performance advantage provided by valuable brands is a significant and robust result.

Within the same line of research, Hendricks and Singhal (1997) find support, in a sample of 101 US publicly traded firms over the period 1984–91, for the hypothesis that delays in new product introduction reduce the market value of firms. Also, they found that more diversified firms experience a less negative impact from being late to the market, other factors being equal. Chen et al. (2002) analyse this issue in a strategic management framework in which the impact of the introduction of a new product on a firm's value depends on the reaction of its rivals. In particular, they distinguish between a rival's passive strategies, that is, stay put, or active strategies, that is, react and imitate. The first case is defined as competition in a 'strategic substitute' framework, whereas the second is defined as competition in 'strategic complements', following Sundaram et al. (1996). In the former case one could expect a positive impact on the share price of the firm that announces the new product and a negative one on its rivals. In the second case, the impacts are ambiguous, as the final effect depends on the interaction between positive and negative effects. The positive effect reflects the gain in competitive advantage derived from the introduction of a new product, whereas the negative one depends on the costs of developing and introducing the new products that imitators do not encounter. Also, a product innovator can additionally face a greater risk and higher probability of mispricing the new product. Chen et al. (2002) use a sample of 384 new product announcements made by 101 different firms for thirty-nine industries between 1991 and 1995. According to them, firms competing in a strategic substitute framework find a positive impact of product innovation on the firm's valuation. On the contrary, the effect is negative if a firm operates in a strategic complements framework, which suggests that the characteristics of competitive interaction in an industry are crucial in assessing the impact of new product introduction.

10.3.3 *Productivity, product innovation, and R&D*

The first of the non-profit-related measures of performance that we have suggested is productivity or the rate of growth of productivity. There is not a large literature that relates productivity to product innovation activity directly, but there is a considerable economic literature that has investigated the impact of R&D spending on productivity growth. Our interest in the R&D–productivity link is justified here on the ground that firms partly undertake R&D investments to create new goods and services (i.e. undertake product innovation) and these may increase the demand for firm's products and thus its sales or prices; although of course new process innovations that result from R&D may reduce the costs of production of existing goods.

Hall et al. (2010) provide a review of the literature that has attempted to estimate a relationship between firm-level R&D and productivity. Estimates of R&D elasticities, based on both levels and growth rates of the variables, range from 0.01 to 0.25, with a central tendency around 0.08. Such studies have generally assumed a Cobb–Douglas production function in which a measure of knowledge stock at the firm level is assumed as a general proxy for various aspects of a firm's innovative attitude. The econometric approach typically used rests on the underlying assumption that R&D investments generate a stock of knowledge at the firm level that may yield future returns.

Although most of the empirical investigations using R&D flows or stocks as proxies for innovation input have proved the existence of a clear relationship between R&D and productivity, these studies fail to disentangle the R&D returns associated with the various type of innovation outcome. For example, it has been argued that R&D associated with product innovation is characterized by a lower rate of return than process innovation (Scherer, 1982). Reasons may be related to the fact that the introduction of new products is generally associated with adjustment costs that may negatively impact on firm productivity in the short run; in addition, the impact on productivity may be underestimated because the effect of quality change on price is poorly measured.

In the last twenty years, due to the availability of community innovation survey (CIS) data, more direct measures of innovation activity have also been employed. It has been possible to disentangle innovation inputs, such as expenditures on various types of innovation investments, and innovation output, such as product or process innovation dummies and share of sales due to new products. An important contribution to the empirical debate is to be found in CDM (1998). This work employs for the first time data on the French CIS, opening the way to a more intensive use of innovation survey microdata. By adopting a Cobb–Douglas production function framework, CDM (1998) derive a simultaneous equations model that links together productivity, innovation output, and R&D spending in a cross-section of industrial firms. They find that innovation output, as proxied either by the share of innovative sales or by patent counts, is positively and strongly affected by R&D.

Since the original CDM (1998) analysis, many empirical studies have applied this simultaneous equation approach to modelling the impact of innovation on productivity. The general model consists of three sets of relationships. In the first set, one or more relationships describe a firm's decision to undertake R&D

and, if it does so, the proportion of resources spent is defined as a function of firm and industry characteristics. The second set describes the various types of innovation outcomes by using probit equations for estimating the probability of innovation as a function of R&D intensity and other firm and industry characteristics. Also, CDM (1998) include an equation for the share of innovative sales (typically the sales share of products introduced during the past three years). The third relationship summarizes the contribution to productivity of expected innovation that is conditional on R&D and other firm characteristics.

Hall (2011) provides a summary of the studies based on the CDM (1998) model or its variants, which have estimated the contribution of product innovation to firm-level productivity. The elasticities on the share of innovative sales range from 0.04 to 0.29, higher values characterizing R&D-intensive and high-tech firms. When product innovation is proxied by a dummy variable, the estimates support a positive relationship indicated by the findings on innovative sales, although somewhat less precisely estimated.

A recent and somewhat more structural modelling approach relies on a model specification that tries to disentangle the effects of product and process innovation by embodying both productivity effects and demand shifts. In Jaumandreu and Mairesse (2010, 2016), for a sample of Spanish firms, product and process innovations are represented by binary indicators and are specified as demand shifters. Results support the view that product innovation increases demand, with an average effect of 3.3 per cent. The impact of process innovation, when associated with product innovation, appears to be more controversial, as it increases marginal costs in five out of ten industries. Using a similar framework, Peters et al. (2016) estimate the impact of R&D investment on the long-run return for a panel of German companies, underlining how the firms' financial strength affects or plays a role in their performance. The estimated expected revenue increase is industry-related and varies from 9.3 per cent in electronics to more than 16 per cent in vehicles. Such higher returns bring about higher probabilities of both process and product innovation. The latter occurs more frequently within firms that show a stronger financial strength.

Related work by Petrin and Warzynski (2012) develops a methodology for estimating demand and production functions and attempts to determine the effect of firm-level R&D on product quality and production efficiency. In their model, R&D affects product quality, which is a function of past product quality, technical efficiency, past R&D investments, and possibly other control variables. Results support the view that product quality is strongly negatively correlated with technical efficiency, which is interpreted as a possible trade-off between product and process innovations. R&D expenditure as a share of firms' sales positively affects both product quality and technical efficiency, but, when R&D expenditure is disentangled between product and process innovations, the former significantly predict quality improvements but not technical efficiency. Conversely, R&D expenditure on process innovations is positively associated with technical efficiency, whereas the effect on quality improvements is more controversial. All in all, these findings indicate the need for a better understanding of the effects of different mixes of product and process innovation in structural equations modelling.

Another relevant issue is that of causality between innovation and productivity. More recent literature has increasingly focused on this topic, partly because of the availability of larger panel data sets. This issue is relevant because, although empirical literature has proven the relevance of innovation for firm productivity, there is not a well-established understanding of the causal link between these variables. In other words, one may ask whether more efficient firms invest more in R&D, or whether firms investing a great proportion of their resources in R&D experience higher productivity. Also, simultaneous causality may be the prevailing pattern, thus ruling out clear-cut causation directions. Among the most significant contributions, Rouvinen (2002) analysed Granger causality between R&D and productivity by using an unbalanced panel covering fourteen industries in twelve OECD countries over the period 1973–97. He found that R&D does cause productivity (but not vice-versa) and that the impact of R&D varies in timing and magnitude. Frantzen (2003) found that causality runs mainly from R&D to productivity; he found this by using a panel of twenty-two manufacturing industries and a vector autoregression (VAR) methodology in fourteen OECD countries over the period 1972–94. Battisti, et al. (2010) explored Granger causality between R&D and productivity within a balanced panel of 552 UK firms over twelve years by using a generalized method of moments approach. They suggest that R&D does cause productivity, whereas productivity has no influence on R&D. Also, by using a non-parametric approach—namely data envelopment analysis (DEA)—they attempt to test the role of unobserved individual heterogeneity, which has been modelled in a longitudinal context as firm fixed effects, together with R&D spending as causal factors of productivity growth. They found that firms that have both high R&D spending and high firm fixed effects do reach higher productivity growth scores, but high R&D spending or firm fixed effects alone do not imply high productivity growth.

10.3.4 *Product innovation and exporting*

The existence of a relationship between product innovation and a firm's export performance has been suggested by different theoretical models. In the early product life cycle models discussed by Vernon (1966), product innovation is an exogenous determinant of a country's exports. On the other hand, endogenous growth theory has explored the endogenous determination of innovation (Grossman and Helpman, 1991), which suggests the possibility of a reverse causal direction running from exporting to innovation and capturing two different mechanisms. The first reflects the fact that, when operating in international markets, companies need to invest further in R&D to stimulate both product and process innovation and thus to maintain competitive advantages. The second reflects the learning-by-exporting effect, as exporting firms have better access to new knowledge and technological competencies than firms operating only in the domestic market and thus may have a higher propensity to introduce new products.

More recently, a central issue in the theoretical debate has been the role of firms' heterogeneity in explaining the pattern of internationalization. The hypothesis underlying this approach is that exporting firms exhibit specific characteristics, in particular higher productivity levels that enable them to bear the large sunk costs required to compete in international markets (Melitz, 2003). However, most of these models assume productivity differences as exogenous, thus failing to describe potential linkages between firm productivity, export decisions, and innovation.

Pursuing a similar modelling framework, Bustos (2010) and Caldera (2010) develop a theoretical approach where productivity differentials among firms are not randomly determined but arise because of different patterns of technology adoption. Their findings suggest that innovative firms have a higher propensity to export, both because of the expected higher revenues generated from product upgrading and because of the lower marginal costs of production, which allow them to reduce selling prices (Bustos, 2010). Caldera (2010) also finds that product upgrading has a greater impact than cost reduction (i.e. process innovation) on the propensity to export among Spanish manufacturing firms. Although explicitly considered in her model, product innovation is interpreted, like in Bustos (2010), as a cost-reducing innovation that allows a company to charge a lower price. However, the effect of quality improvement on export performance in not considered.

Attempts to endogenize quality improvements have been developed by Borin (2008) and Antoniades (2015). In their models the firm must decide whether to increase its mark-up for higher quality products. Both papers conclude that exporting firms are more productive than non-exporters. Also, as the most productive firms tend to produce higher quality products, on average companies sell qualitatively superior goods on international markets.

Some firm-level empirical investigations have analysed the effects of product innovation on exporting by using survey-based measures of innovation. Many of the early studies are cross-sectional and consider innovation as an exogenous variable with mixed results (Wakelin, 1998; Bernard and Jensen, 1999; Sterlacchini, 1999). Roper and Love (2002) provide a comparative analysis using plant-level data from the United Kingdom and Germany. Export propensity, measured as the ratio of export value to total sales, is regressed on three different innovation indicators: (i) a binary product innovation index; (ii) a measure of innovation intensity, defined as the number of product changes at the plant level per employee; and (iii) a measure of innovation success defined as the share of innovative sales at the plant level. Results are quite controversial as, although the decision to introduce product innovation positively affects the export probability, export propensity, that is, the proportion of a plant's sales that are due to export, is not significantly affected. In addition, innovation intensity does not affect export probability or export propensity in either country. Finally, innovation success has a positive and significant effect on export propensity in the United Kingdom, while the effect is not significant in Germany. However, it is worth noting that these results are not derived from a true panel data set and therefore cannot be interpreted unambiguously.

The relationship between product innovation and exporting is analysed by Ganotakis and Love (2011) by using a cross-section of UK new technology-based firms (NTBFs). In their model endogeneity between product innovation and export is addressed through instrumental variable techniques. The model also controls for possible self-selection among the exporters (given that the selected sample does not derive from a random sampling of all NTBFs). Results indicate that NTBFs introducing a new-to-market product innovation show an export probability that is 40 per cent higher than that of non-innovating firms when endogeneity is specifically addressed. When adopting a measure of innovation performance (proportion of innovative sales), this result is confirmed, with a marginal effect that is significantly positive although very small in size (0.004).

Becker and Egger (2013) adopt a propensity score matching (PSM) approach to control for sample selection bias. PSM is a statistical technique used to minimize the effects of confounding when estimating treatment effects using observable (non-randomized) data (Rosenbaum and Rubin, 1985). Also, to take into account the endogeneity of product and process innovation, when estimating the impact on export propensity, these authors resort to a bivariate probit model, using past values of both product and process innovation together with other contemporaneous and lagged regressors. Using a sample of German firms observed over a period of three years and available from the Institute for Economic Research (IFO), they find that product innovation is more relevant than process innovation to the firm's export decision. The estimated probability of exporting for firms that introduce only product innovation is about 16 percentage points (p.p.) higher than that for firms that adopt only process innovation and 7.4 p.p. higher than that for non-innovators; that is, the predicted probability of exporting for firms that introduce only process innovation is lower both than that of product innovators and than that of non-innovators.

Product innovation is viewed by Cassiman et al. (2010) as a determinant of small firms' decisions to export. Using a panel of small and medium-sized

Spanish businesses and a total-factor productivity (TFP) index as a proxy for firms' productivity, they compare the productivity distributions of exporters and non-exporters by using a Kolmogorov–Smirnov equality-of-distribution test. Results indicate that the most productive firms tend to export more. In addition, they use a transition probability approach to relate exports to prior innovation decisions. The authors find support for the hypothesis that product innovation is a key driver of exports, as it increases the probability of switching from not being an exporter to being an exporter by 49 per cent, by comparison with 36 per cent for firms that introduce only process innovations. The reverse is also true, that is, the probability of moving from being an exporter to being a non-exporter is lower for firms introducing a product innovation than it is for process innovators.

While the study by Cassiman et al. (2010) finds support for the existence of a relationship running from product innovation to exporting activity, Damijan et al. (2010) investigate the reverse causation, that is, the link running from exporting to innovation, and also to productivity growth. By using an integrated database of Slovenian firms that links three successive CIS surveys to economic and financial data and foreign trade flows, they combine a PSM approach with a transition probability analysis to estimate the likelihood of becoming an exporter. They show that the decision to export does not significantly affect product innovation. They test this hypothesis by observing the increase in the number of goods sold during a three-year time span after starting to export. Conversely, they proxy the possible effect of the decision to export on process innovation by computing the impact of exporting on the change in TFP. The observed impact is positive and significant over the entire period, thus suggesting that operating in an international market positively affects firm efficiency by stimulating process innovations; and this supports the learning-by-exporting hypothesis. They do not find evidence that product or process innovation increases the probability of exporting.

Following the analyses of Melitz (2003), several studies have specifically addressed the issue of changes in product quality and of their impact upon exporting activity. Using the tenth wave of the UniCredit Survey of Italian Manufacturing Firms and balance sheet information derived from the AIDA database, Imbriani et al. (2015) aim to test a theoretical model where firms set their prices as a markup over the marginal cost, thus allowing for quality improvement. A measure of quality improvement is derived as a dummy variable indicating whether firms have invested in a quality upgrading of an existing product. Export propensity is estimated by using a propensity score model that includes among the regressors (i) a quality improvement dummy together with other two dummies that capture product innovation (development of new products) and production cost reduction; (ii) a measure of past TFP; (iii) a measure of past R&D investment; and (iv) a set of financial indicators that capture the extent of financial constraints. Quality improvement (but not, surprisingly, new product introduction) is shown to increase the export probability, although the results are not significant at conventional levels.

Building upon a related trade model, Crozet et al. (2012) reach similar conclusions in an empirical analysis of the Champagne market. The original feature of their approach is that they employ a direct measure of product quality for individual producers that is derived from a guidebook also reporting comparable information on unit values and market shares. Producers are then classified with one to five stars and this information is used as an explanatory variable in a set of three equations for free-on-board (FOB) prices, export propensity, and export performance (log of export revenues). Higher quality tends to increase export prices, export probability, and export performance, with a relationship that tends to be monotonic. The authors' probit estimation implies that, by comparison to lower quality competitors, a five-star producer gains a probability premium that ranges from 14 p.p. to 5 p.p.

10.4 **Innovation complementarities**

Recent theoretical and empirical research has increasingly recognized that looking at the adoption of single innovations or types of innovations may be misleading, since firms often tend to simultaneously undertake groups of innovation activities encompassing not only technological aspects of the productive process but also marketing, organizational, and workforce management strategies (see also chapter 8). Following this line of reasoning, the existence of multiple interactions between technological and non-technological factors of production may give rise to complementarity patterns insofar as doing more of any one subset of a group of innovation activities may increase the returns from doing more of the remaining activities in other subsets. The supposition is that the simultaneous undertaking of several innovative practices is not an accident, but rather the result of coordinated action between traditionally separate activities and work practices. With such a view, Milgrom and Roberts (1990, 1995) explore the case of a multiproduct profit-maximizing firm by formalizing a theoretical model in which the firm chooses a set of decision variables that are supposed to form clusters of complementarities. By using mathematical lattice theory, first introduced by Topkis (1978), they argue that there may be extensive complementarities within manufacturing firms that cause them to adopt a strategic clustering approach to innovation.

The empirical literature on complementarities is extensive. A review of the management research may be found in Ennen and Richter (2010). Within economics a growing number of scholars have emphasized the complementarity between different aspects of innovation, for example technological and non-technological innovations, yielding a stream of studies that use microdata derived from innovation surveys. According to CIS definitions (OECD, 2005), non-technological innovation includes marketing and organizational innovation, where the former is defined as 'the implementation of a new marketing method involving significant changes in product design or packaging, product placement, product promotion or pricing' and the latter as 'the implementation of a new organizational method in the firm's practices, workplace organisation or external relations'.

Following the approach suggested by Milgrom and Roberts (1990), testing for innovation complementarities within CIS-type microdata involves regressing a firm's performance variable on a set of control variables (dummies), which reflect the innovation status under investigation. Thus a joint test of different alternatives enables one to ascertain whether complementarity prevails or not, provided that the equation has been correctly specified. This is the approach originally proposed by Mohnen and Roller (2005) and also applied by Doran (2012) and Ballot et al. (2015). The former study concentrates on a set of complementarities related to new-to-market or new-to-firm process or product innovations, whereas the latter analyses complementarities between product, process, and organization innovation in a comparative framework. Complementarities are shown to exist.

Earlier Schmidt and Rammer (2007) and Schubert (2010) used the German CIS to test whether marketing and organizational innovation are complements to or substitutes for product or process innovation. Both analyses were performed on a cross-sectional basis, Schubert (2010) using data from the 2007 wave of the German CIS and Schmidt and Rammer (2007) using the 2005 data. In both cases the authors found that, when focusing on a large set of manufacturing and service firms that participated in these surveys, marketing and organizational innovation does complement technological innovation, which implies a positive impact on firms' performance. However, this impact crucially depends on the performance measure that is adopted. Indeed, Schubert found that the percentage of sales due to new products—a measure of innovation success—increases and costs are reduced when marketing innovation is simultaneously introduced with product or process innovation. These findings confirm the results of Schmidt and Rammer (2007), who also found a significant effect of both marketing and organizational innovation on innovative sales and cost reductions for those firms that also introduced product and process innovation. They also tested for the impact on profit margins—a measure of the economic success of a firm—by using estimates that are related to ordinal measures of the profit variable (ordered probit estimation). However, using this approach, they find that the largest effect on profit margins is attributable to technological innovation alone, suggesting that non-technological innovation may not be relevant for the economic success of the firm.

The relationships between marketing innovation and innovation performance are explored in a dynamic context by Lhuillery (2014). Lhuillery used an unbalanced panel of manufacturing firms, which he obtained by matching four consecutive waves of the French CIS. Sales of new or improved products are used as an indicator of innovation success. He found that marketing innovation has a short-term direct (contemporaneous) effect on innovation success, whereas the long-term (lagged) effect is not significant. In high-tech sectors the short-term effect is not significant for incremental products. In order to test for the role of marketing in enhancing the persistence of innovation success, an interactive term between the lagged share of innovative sales and a lagged dummy for marketing innovation was also introduced. Results support the view that marketing innovation does not enhance the persistence of product innovation in low-tech industries. In high-tech industries results are more controversial, as the interactive coefficient is positive and significant for incremental innovation, whereas it is negative and significant for radical innovation.

Bartoloni and Baussola (2015) use an unbalanced panel of Italian manufacturing firms to investigate empirically the impact of product innovation on firms' economic performance, pinpointing complementarities between product and marketing innovation during the period 1998–2008. By using a simultaneous equation approach to the estimation of a firm's profitability and productivity, they find that product innovation, when performed persistently and in conjunction with marketing activities, significantly increases profitability.

Battisti and Stoneman (2010) used the fourth UK CIS to explore the impact of the adoption of a range of innovative activities, such as product, machinery, marketing, organization, management, and strategic innovations. By using a clustering approach, they show that there is a significant degree of complementarity between these innovation practices. They identify two major sets of innovations: on the one hand, marketing, organization, management, and strategic innovations (labelled wide innovation) and, on the other, more traditional activities: machinery, process and product innovations (labelled technological innovation). Wide organizational innovation is found to play a crucial role in the innovative activity of UK firms. The authors find a positive impact of such activities on the firms' performance; however, they do so by using a qualitative and subjective measure of performance derived from their respondents' judgments on the impact on future value added—a data-imposed limitation.

Battisti et al. (2005) take particular care to model directions of causation but still find the existence of extra profit gains from the joint rather than the individual adoption of different work practices. Such innovations may thus be complements, that is, the overall net gain from their joint adoption is higher than the sum of the net gains from individual adoption. Battisti and Iona

(2009), Ichniowski, Shaw, and Prennushi (1997), and Whittington et al. (1999) offer other examples of super-additivity in clusters of innovations, while the formalized models of Battisti et al. (2005) allow in principle for both substitute and complementary technologies.

10.5 Product portfolios, firm performance, and the product life cycle

At any point in time a number of different innovation projects may be present within a company's investment portfolio and encompass a range from basic ideas still in the exploration stage through to the development of incremental or radical product innovation. In such circumstances a single overall measure of a firm's economic performance may not accurately reflect the performance of all the innovative efforts that are simultaneously being undertaken. Different types of innovation may require the use of different metrics.

For example, incremental innovation, which provides small but continuous changes in products that help to improve product performance, is, for most companies, the dominant form of innovation. For such incremental innovation, economic performance may be relatively easy to measure by using conventional financial metrics. On the other hand, successful radical innovation occurs much more infrequently, and the risk of failure may be very high.[5] Measuring the final impact of such innovation on a firm's performance is much more difficult, because (i) the associated financial risk is relevant and (ii) capturing the accumulated cash flow generated over the whole life of a radical product innovation may be particularly critical, given that most radical innovations take considerable time to become accepted. In addition, the development of a radical innovation may generate organizational capabilities and learning processes that can be extended to further innovative efforts. In this context, it is hard to disentangle how much of the cash flow is attributable to the first radical innovation and how much to the other innovative projects.

The PLC approach (Klepper, 1996) may be seen as an attempt to characterize the typical cash-flow profile of an innovation project over time. This approach reflects that a successful product innovation will take time to return the investment in its generation but, once launched on the market, cumulative cash flow will initially grow fast and then will gradually slow down as the product ages, until it no longer increases significantly. Behind this story is the view that, according to the Schumpeterian view of competition (Schumpeter,

[5] According to Larry Keely, President of the Doblin Group, 96 per cent of radical innovation fails (*Bloomberg Business Week*, 2005, August 1: 'Get creative: How to build effective companies').

1934), firms engage in risky innovative efforts when they foresee prospects for gaining competitive advantages by creating products or services that are preferred by the market. The launch of the product is the first marker of this comparative advantage, but then a non-linear process involving the entry and exit of firms into the given industry or market causes evolutionary changes over time in the products offered and sold on the market and in the profits realized. This pattern (also referred to in chapter 4, section 5) is often known as a product life cycle. The approach suggests that profitability is affected by product innovation because firms compete to attract new buyers, reducing the ability of previous innovators to acquire new customers and therefore reducing expected returns from product innovation. However, product innovation may decline over time, due to firms' exiting the market. According to this view, innovation is strictly related to firm size and is proportional to a firm's level of output. Thus the incentive to introduce either a product or a process innovation crucially depends on the industry life cycle and on the ability of the innovating firms to exploit their technological opportunities. In such a context, firms have an incentive to pursue product innovation at an early stage, whereas, as an industry matures, process innovation becomes more attractive.

The PLC approach has been challenged on different grounds, in particular by management studies that have emphasized how a product may be repositioned by a new company strategy. Moon (2005) suggests, by using appropriate case studies,[6] that the PLC hypothesis may be revisited in the light of strategic refinements that enable firms to exploit a product's potential even if the product itself may be categorized as mature. However, it is worth underlining that case studies must be evaluated over a longer period, to prove a clear-cut and robust evidence of the relationships under investigation. Indeed, some of the author's cases did not prove significant in later periods.

In fact managerial or strategic behaviour such as portfolio management, that is, the dynamic decision-making process by which the firm's active portfolio of new product projects is continuously updated and revised, is argued by some to be critical to a firm's performance (Lofsten, 2014; Cooper, Edgett, and Kleinschmidt, 2001). For example, Pisano (2015) proposes an interpretative matrix that summarizes the challenges faced by companies when they have to define an innovation strategy. The firm's innovation landscape is grouped into four categories: (i) disruptive; (ii) architectural; (iii) routine; and (iv) radical. The firm must select a strategy, that is, a location in this matrix, and allocate resources accordingly, taking into account the innovative characteristics of its product offerings. It is worth noting that disruptive, architectural, and radical are not necessarily the most profitable strategies. In reality, routine innovation is extremely relevant in terms of

[6] In particular, Moon refers to reverse positioning, using IKEA as an example and citing Heinz EZ Squirt as a case of breakaway by a packaged good.

profits and its role cannot be undermined. Such decisions will be industry- and firm-specific in that the firm has to take into account sectoral technological opportunities, its and other firms' market power, and consumer needs and how they can be met.

Another, complementary, approach is proposed by Davis (2002), who offers a selective methodology embodied in a matrix that indicates how the impact on net present value adjusted for risk (NPVR) of a product innovation strategy may be considered in terms of a trade-off between market risk and product risk. He divides the typical product portfolio categories (new venture, new platform, and new category-derivative product) into a risk-weighted scheme. In particular, one can think of the traditional product portfolio categories that are then split into four risk categories. Such an approach would enable a firm to calculate its NPVR, which may be compared with the simple NPV, thus providing a measure of the projected risk of its strategy.

10.6 **Product differentiation and firm performance**

The strategic management literature suggests two alternative approaches to achieving competitive advantages: product differentiation and cost leadership (Porter, 1980; Hall, 1980). Product differentiation and product innovation are close relatives. Although differentiation may be defined by multiple attributes, in this product differentiation literature it is usually implicitly assumed that the differentiation is vertical, that is, all consumers have the same ranking order for the products on the market. Given that, for a given price, consumers prefer better quality, all else being equal, the literature typically refers to differentiation as reflecting differences in an agreed quality dimension. In contrast, in a horizontally differentiated product space, consumers rank product characteristics differently and different consumers have different preference rankings of a product's variants. If all the variants of a product are sold at the same price, different buyers will purchase different products.

With vertically differentiating product innovation, conventional wisdom has suggested the existence of a trade-off between quality and cost, in that higher quality usually requires more sophisticated production techniques, less standardized materials, and the complementary use of other activities (for example, marketing functions) that are incompatible with low costs. The effect of vertical differentiation on firm performance is assumed to be positive. Empirical evidence shows that product quality positively influences market share (Buzzell and Wiersema 1981a, 1981b; Flaherty 1981) and also may reduce direct costs due to experience-based cost saving.

By adopting a trade model of monopolistic competition and a firm-level export data set on champagne producers, Crozet et al. (2012) show that quality determines the prices that firms charge, the set of countries to which their exports are directed, and the amount of exports. The results, therefore, show that higher quality has a positive impact on firm performance and on success in foreign markets. By using a customer satisfaction approach, Cadotte, Woodruff, and Jenkins (1987) suggest that higher product quality may be positively associated with higher profitability, because superior quality ensures that customers will repurchase the same product. In contrast, low product quality may have a detrimental effect on repurchase decisions (Anderson and Sullivan, 1993), thus reducing firm profitability. The empirical literature has also addressed the supposed incompatibility of high quality and cost leadership (Reitsperger et al., 1993; Jing, 2006). Phillips et al. (1983) conclude, on the basis of empirical work on to a sample of manufacturing firms, that quality and cost control interact to generate above-average ROI.

Although the economic literature has long recognized the distinction between vertical and horizontal differentiation (d'Aspremont et al., 1979; Shaked and Sutton, 1987; Hotelling, 1929), there are very few empirical results that inform as to whether the impacts on firm performance differ in type or degree. Theoretically there is greater insight. The theoretical models by Neven and Thisse (1989) and Economides (1989) analyse two-dimensional vertical and horizontal differentiation cases in which firms compete on quality and variety, as well as on price. These models, however, reach quite different conclusions: Economides' (1989) framework leads to maximum horizontal differentiation and a minimum vertical differentiation, while Neven and Thisse (1989) show that firms choose a maximal differentiation in either the horizontal or the vertical dimension, with minimum differentiation in the other. The conjoint effect of horizontal and vertical differentiation has also been analysed by Ferreira and Thisse (1996), who constructed a two-stage game model. They conclude that the subgame perfect Nash equilibria involve minimum (maximum) vertical product differentiation when horizontal product differentiation is sufficiently large (small).

Within the literature on imitation, multidimensional product differentiation is viewed by Ethiraj and Zhu (2008) as a strategic tool for increasing the imitator's performance, as measured by the probability of eroding the innovator's sales advantage. The results (which relate to the case of imitation in the branded drug industry), support the view that a differentiation strategy is conditioned by the amount and quality of information available about the innovator's product. Thus, when uncertainty is high, horizontal differentiation may increase the likelihood of success, whereas vertical differentiation seems more appropriate when uncertainty is low. Jang and Park (2012) use panel data on a sample of Korean investment or brokerage firms to investigate the effects of the interaction between vertical and horizontal differentiation on the firm's performance. They use the total time spent by visitors on a website during a specified period as a proxy for the performance of an online trading system, by assuming a positive correlation between the total time spent and the online brokerage fees. The availability of an external evaluation system for the online trading market allows for the development of differentiation measures that incorporate both quality and variety attributes. Results support the view that a mixed effect is at work, horizontal differentiation having a smaller effect on a firm's performance when the overall quality is low.

Product differentiation has been particularly widely addressed within the marketing literature, where the focus is primarily on how product differentiation, as reflected in branding, impinges on firm performance. Examples of horizontal and vertical differentiation may be so-called line and brand extensions. The former type encompasses an expansion of the existing product line by modifying a specific characteristic (e.g., diversified soup tastes from a soup producers, or customized designed seats for a car model) while the latter type refers to an extension of the brand to a new product that involves either introducing a vertically differentiated product or a new product in a new category, thus using the brand as a driver for the success of the new product.

Related empirical research has mainly focused on issues that are typically linked to marketing strategies; but the research that is most relevant to the topic here has focused on brand value and its impact on firms' performance, for example a firm's market value. One study that does not explicitly consider firms' financial indexes also indicates that brands can have a positive value. Crass (2014) investigates the effect of brand use on innovation performance employing data from the German 2011 CIS. Results suggest that branded product innovations are more successful in generating sales of new products when the firm decides to use an established brand. Conversely, the effect is not significant when the market introduction of the new product is under a new brand.

10.7 Market orientation and the product development process

In recent years, characterized as they are by high competition in global marketplaces, the ability to cope with customers' needs is a core strategic issue, but firms may have been facing increasing complexity in the development of new products that should meet customers' requirements. As a result, an increasing number of scholars have focused on the notion of firm's 'market orientation' in product development with the aim of understanding its link to firm performance. Within the management science literature, market orientation has been defined as a form of organizational culture (Deshpande and Webster Jr, 1989; Narver and Slater, 1990; Day, 1994). In the definition of

Narver and Slater (1990), a market-oriented firm is one that manifests a customer and a competitor orientation, together with inter-functional coordination. It has been argued that market orientation, when combined with organizational capabilities and learning orientation, may increase a firm's ability to interpret customers' needs, and thus to innovate successfully (Hurley and Hult, 1998; Gatignon and Xuereb, 1997). The strategic role of managerial competencies in enhancing a firm's profitability is also emphasized in the resource-based view of the firm (Penrose, 1959; Wernerfelt, 1984) and, more specifically, in the dynamic capabilities approach (Teece, Tsano and Shuen, 1997). According to this latter view, the firm achieves competitive advantages by organizational improvements and learning processes, in order to adapt to a continuously changing business environment.

Studies appearing in the organization-oriented tradition have tried to conceptualize and then test the role of market orientation in creating improved organizational performance. These studies are based on appropriate surveys and on the use of ad hoc variables indicating organizational culture (Hurley and Hult, 1998) or, more specifically, marketing orientation (Narver and Slater, 1990; Slater and Narver, 1994; Atuahene-Gima, 1996; Han et al., 1998). Potential limitations of this stream of empirical research may be found in the extensive use of subjective measures (mainly on a cross-sectional basis); these concerns are clearly recognized by the scholars in the field of organization studies. Slater and Narver (1994) recommend using different sources of data and the introduction of objective measures of firm performance. However, in the organization and management science literature, Narver and Slater (1990), through the use of ordinary least squares regressions on a sample covering 140 strategic business units, showed that market orientation and performance, as measured by relative return on investments, ROA, are strongly related, thus suggesting that market orientation is one driver of a firm's competitive advantage strategy. The model controls for firm-specific characteristics (relative costs and relative size) and for market-level factors (growth, concentration, entry barriers, buyer power, seller power, and technological change). Narver and Slater's original results have been further confirmed (Slater and Narver, 1994) insofar as market orientation does have a long-term impact on business performance, whereas environmental conditions (e.g., a competitive environment) only have short-term effects.

These findings opened the way to other studies that have refined the original conceptual framework. Using multivariate tests of significance on data from forty-five product development projects in twelve firms, Olson et al. (1995) found a positive impact, on a firm's performance, of the availability of coordinated functional departments. This evidence implies that marketing and R&D department integration impacts positively on firm performance. Further support for the complementarity role of market orientation in the product development process arises from the Han et al.'s (1998) study, which focused specifically on market orientation, according to the Narver and Slater's (1990) definition. Using a three-stage least squares estimation technique on a sample of 225 banks, Han et al. found a positive and significant impact of market orientation (i) on innovation (defined as technical and administrative innovations) and (ii) on business performance (measured in income growth and return on assets), this relationship being mediated by innovation. This finding supports the view that market orientation is an important aspect of innovation activity for achieving superior performance.

The view that market orientation and innovation are complementary in generating performance characteristics is also supported by Atuahene-Gima (1996). On the basis of a sample of 275 firms from both the manufacturing and services sectors, he found that market orientation contributes to innovation performance, measured both in terms of 'market success' (a self-reported measure of sales, market share, and profits) and in terms of 'project impact performance' (a measure of cost efficiency). Interestingly, market orientation has a stronger effect on internal efficiency than on market success. However, this result seems to contrast with the analysis of Baker and Sinkula (2009). which, based on a structural equation-modelling approach applied to a small sample of eighty-eight firms, provides support for a direct effect on profitability (in terms of self-reported changes in sales revenues, profit, and profit margins) of market orientation (via a modified version of the Narver and Slater's scale), but not of entrepreneurial orientation (which included firm's innovativeness as an input measure of innovation). They also found a positive and significant effect of a measure of successful innovation on profitability, thus indicating that an output rather than an input measure of innovation should be preferred. Gatignon and Xuereb (1997) include technological and organizational issues within the Narver and Slater approach. They emphasize how firms must be consumer- and technology-oriented in those markets characterized by high demand uncertainty in order to be able to market innovations (new products). Conversely, when markets are less turbulent and thus demand is relatively stable, a competitive orientation is more relevant for marketing innovation.

10.8 Conclusions

In this chapter we have explored the impact of product innovation on firm performance, encompassing several different approaches found in the literature. In particular, we have attempted to address both economic and managerial literatures, although with some emphasis upon the former.

On the one hand, the economic literature emphasizes quantitative measures of the impact of product innovation by concentrating upon accounting indicators, market-based measures of firm performance, or other explicitly economic indicators such as productivity or export activity. On the other hand, the managerial approach, in addition to considering other quantitative measures, also emphasizes the role of the innovation process that leads up to product innovation, as well as the innovation itself, in the determination of firms' performance. Typically, the managerial approach refers to quantitative measures based on cost accounting—measures that enable managers to rank innovation projects according to their expected net benefits. The managerial approach also implies that to study the impact of product innovation on firm performance requires the analysis of the firm's strategy towards product innovation, its business model, and its product portfolio (among other factors). It has also been argued in this literature that such analysis should be considered within the context of the PLC in order to provide a dynamic view of the impact on firms' performance.

The impact of product innovation on firms' profitability has been extensively discussed in the economic and managerial literature. The most relevant finding in the current context is that there is strong support for the view that product innovation has positive and significant short- and long-run effects on firm profitability. This impact varies across industries but seems independent of the number of innovations introduced in each sector. In addition, the role of complementarities in improving firms' performance has been stressed. In particular, marketing and organizational innovations may complement product innovations, thus enabling companies to gain both profit and productivity premium. Additionally, the impact of product innovation may be estimated indirectly, by viewing firms as bundles of tangible and intangible assets whose values are determined by the financial markets. The impact of R&D and patents (to which product innovation is closely related) on firm market value may be evaluated using this analytical framework. Hall (1999) provides an exhaustive synthesis of the main results derived from international comparisons, although it is worth underlining that studies for the United States and United Kingdom prevail, given that larger data sets for listed companies are more available in these contexts than, say, in continental Europe. The impact of R&D expenditure on firms' market value ranges from 2.5 per cent to 8 per cent. Additional evidence for some European countries (e.g. France and Germany) shows coefficients on the R&D stock (respectively 0.28 and 0.33) that are lower by comparison with the United Kingdom and the United States.

In the absence of much literature relating directly to the impact of product innovation, we have discussed the results of empirical analyses that deal with the impact of R&D and patents on productivity. Many recent empirical studies have taken advantage of the availability of panel data incorporating innovation variables. A significant contribution on methodological grounds is provided by CDM (1998), who suggest a simultaneous equation model for testing the relationship between innovation and productivity at the firm level. Hall (2011) provides a comprehensive summary of the main results derived from the empirical literature. The elasticity of the firm's productivity to its share of innovative sales ranges from 0.04 to 0.29, with higher values for R&D-intensive and higher tech firms.

The extent to which, and the success with which, firms compete in foreign markets is also found to be positively related to product innovation and firms' performance on the whole. The empirical literature, using both econometric and PSM models, suggests that firms' export propensities are positively impacted by product innovation and product quality improvements. This relationship should be considered in a framework in which exports are affected by firms' productivity, which in turn is related to those firms' technological capabilities, including their product-innovating activity. However, it is worth noting that causation is rarely analysed deeply and one could think of a reverse relationship, which views exports as knowledge acquired from abroad and enabling a firm to learn and improve its products. There is therefore room for more detailed investigations that may eventually take advantage of longitudinal data in the future.

Product differentiation, either horizontal or vertical, is relevant in determining a firm's innovation strategy and performance. However, there are only a few empirical analyses that explicitly consider the impact of product differentiation on firms' performance. This is probably because most of the attention has been devoted to the development of theoretical strategic models on the economic side, and on the managerial side this aspect has been largely confined to the marketing literature dealing with brands and so-called line or brand extensions. We have therefore considered this line of empirical research mainly by focusing on results that have highlighted the impact of the firms' brand value on their stock valuation (and ultimately shareholder value). The brand value effect is confirmed by empirical work that explicitly relies on financial performance measures.

The impact of new product announcements may also be viewed as a strategic tool that enables companies to gain competitive advantage. This line of empirical research uses the traditional CAPM to derive companies' excess returns and then their relationship with new product announcement and introduction. The evidence confirms a positive and significant impact of information concerning new products on a company's market value, with a higher impact for technologically based industries.

11 Product Innovation and Price Measurement

11.1 Introduction

If one is to obtain an accurate measure of real incomes in an economy over time, or measure changes in output levels, at either the macroeconomic or the microeconomic level, it is important that one has relevant price indicators by which to deflate the nominal (i.e. market price) measures collected by statistical surveys. Except in the rare case where a product innovation is completely original (and thus there is no comparator), new products introduced on to a market may (i) be offered at different prices from those of existing variants and/or (ii) embody quality changes. Depending on whether the new product is an horizontal or a vertical innovation, the quality may be agreed by all to be higher or lower than that of existing products, or the product may be agreed by some to be better than the one previously available and by others to be an inferior product. Such changes in characteristics should, as suggested by Eurostat (2001), be considered to be changes in the volume of output, and not variations in price; and thus a key statistical objective is to generate price indicators that reflect the price of a product of constant quality, that is, to modify nominal prices for an underlying quality change. Such indexes will not only provide a more accurate indicator of inflation, real output, and real incomes but may also, through a comparison with non-adjusted indexes, give some indication of the contribution of product innovations to prices, output, and income growth. The aim of this chapter is to address the main issues that arise in the measurement of price change and then to derive the implications for when the quality of goods changes through product innovation.

11.2 Patterns of product creation and destruction

Price indexes may be calculated at the single-product level (e.g. the price of computers, or the price of cheese) and are also typically constructed as changes in the cost of a basket of goods, the individual prices of which are weighted by market shares. In the presence of product innovations, not only will market prices be changing, but the goods being purchased will also change, the new replacing the old; and, as this happens, the makeup of the basket (weights) should be changing to reflect it. Product innovation thus affects not only prices but also the products on the market that are being purchased. The art of calculating price indicators is to take account of both these changes. Although in chapter 4 we have used a variety of indicators to illustrate that product innovation is extensive, it is useful here to represent some results that reveal intertemporal patterns of product creation and destruction at the economy-wide level—patterns that underline the extent to which the products purchased are changing over time, and thus the need to properly address quality change in measuring price change.

The use of large scanner databases is one method by which product creation (and destruction) and product prices may be tracked and analysed for a reasonably long period following an approach similar to that used to evaluate establishment and labour market data over the business cycle (Blanchard et al., 1990, Davis and Haltiwanger, 1996). Taking such an approach, Broda and Weinstein (2010) analyse product dynamics for a sample of products (ACNielsen SoundScan system) with universal product codes (UPC bar codes) that can be matched to a single firm, enabling the authors to describe product creation and destruction in the US economy. Using this approach, Broda and Weinstein (2010) underline the importance of multi-product firms and the fact that product dynamics can only therefore be completely represented by considering products rather than the company to be the unit of analysis. A distribution of the number of UPCs across firms may be derived, the data indicating that the distribution is skewed, a large number of firms having a small number of products and a small number of firms having a large number of products. Only the smallest firms (sales under $10 million) sell in single brand and product groups, whereas companies with sales larger than $1 billion often market almost 200 brands within more than fifty product groups.

The data indicate that the rates of birth and death of goods are very similar: about 25 per cent of all goods are new in every year, and at the same time about 24 per cent of all goods cease to exist in the following year. Product turnover is concentrated within existing firms and sectors: 92 per cent of product creation and 97 per cent of product destruction take place within existing firms. Also, surviving UPCs are typically larger (measured as a ratio with respect to total expenditure on all products) than the average existing UPC, by a percentage that increases from 23 per cent to 50 per cent, depending on the time span considered (four or nine years). However, the average market share of existing UPCs is decreasing steadily over time, perhaps on the basis that new products introduced during the observed time span, and not considered in the stock of existing products, may capture increasing shares of expenditure on all products.

Broda and Weinstein (2010) find that net product creation appears to be procyclical, whereas product destruction is countercyclical, but with a lower quantitative magnitude. More precisely, product market data suggest that product creation is more relevant than product destruction over the business cycle (contrary to what has been observed for the labour market, where job destruction is more responsive than job creation: see Blanchard et al., 1990 and Davis and Haltiwanger, 1996). Interestingly, Broda and Weinstein (2010) also found that entry and exit in the product market are much higher than in the labour market.

Summarizing, such explorations suggest that the stylized facts of product creation and destruction are as follows: the extent of product creation and destruction is extensive; the pattern of product turnover is largely the result of product creation and destruction by existing firms rather than by new firms; the introductions of new products affect market shares of existing products; and net product creation appears to be procyclical. Because such product dynamics (i.e. creation and destruction) may bring about quality changes, these stylized facts have significant implications for the measurement of prices, which, generally, are not fully captured in official price measurement.

Broda and Weinstein (2010) argue that, by calculating the price of UPCs in different periods and then computing the mean growth rate, it is possible to derive a measure of the price change for all goods. By this route Broda and Weinstein (2010) suggest that ignoring the introduction of new goods and their replacement of the old would lead to an upward bias in the measured price index (the quality bias) of around 0.8 percentage points per year, that is, a cost-of-living index (COLI) that takes product turnover into account is 0.8 percentage points per year lower than a 'fixed goods' price index like the consumer price index (CPI). This implies that the CPI, as usually measured, may overstate inflation in the United States. Thus, if statistical agencies systematically fail to take product dynamics into consideration, there will be a significant discrepancy between official price indexes and a more reliable cost-of-living measure. Moreover, if the introduction of new goods and their replacement of the old is procyclical, as Broda and Weinstein (2010) suggest, then (i) this may increase the volatility of real variables (consumption) over the cycle above that usually measured; and (ii) sampling error may be amplified, given the high degree of price volatility. Over the following sections in this chapter we consider further this issue of the impact of product innovation.

11.3 Product innovation and price indexes

Standard price indexes typically produced internationally by various statistical agencies—for example Eurostat, the US Bureau of Labor Statistics (BLS), and the UK Office of National Statistics (ONS)—calculate the value of a price index, PI, in time t—PI(t)—as the weighted sum of the prices of the various goods (i, i = 1 . . . N) encompassed by the index from time t relative to their price in time zero. The weights will typically be fixed as expenditure shares in the base year (a Laspeyres index) or as such shares in time t (a Paasche index). Typically, in the case of the most commonly employed Laspeyres methodology, this will yield a price index in time t (this index usually being set to 100 in time zero) as

$$PI(t) = \sum_i w_i(0)\ [P_i(t)/P_i(0)]$$

where $w_i(0)$ is the weight assigned to item i according to the expenditure share in the base period 0 and $P_i(t)$ and $P_i(0)$ are the measured (market) prices of the goods in the expenditure bundle in time t and time zero. If the quality of goods does not change over time, variations in this index are attributable exclusively to changes in price and the index will offer an unbiased reflection of how the prices of goods of constant quality are changing over time.

However, if, through product innovation, the products that are included in the expenditure bundle are changing in quality over time, that is, between time zero and time t, then the simple procedure just described will yield a measured price index that no longer reflects the prices of goods of constant quality. In fact the measure will reflect both changes in quality and changes in prices. Thus, whenever quality change is occurring and base period and current period products differ in quality, the price index has to be corrected in some way if that index is to solely reflect changes in prices, holding quality constant.

It becomes crucial to disentangle the price change that one observes with respect to the base period between a pure price change and a price change determined by quality improvements. Without such corrections incorrect measure of product prices may be employed, and this would lead, for example, to biased measure of economic performance (such as inflation or real output). Thus it becomes essential to address the issue of the construction of adequate price indexes that may account for both price and quality change.

One of the most important contributions to the debate concerning the quality adjustment of price indexes is the Boskin Commission Report[1] (Boskin et al., 1996). The Boskin Commission, an advisory commission designed to study the CPI, was appointed by the US Senate Finance Committee in June 1995, with the aim of providing suggestions that would enable the correction of biases in the price index measurement practices of the day. Employing a methodology that enabled the Commission to split the CPI into twenty-seven different categories and then to estimate separately quality change bias for each, the estimate provided by the Commission suggested that

[1] It is worth recalling that other advisory commissions have been established in the United States, e.g. the 1961 Stigler Commission, which underlines how the issue of correctly measuring quality change has been under constant scrutiny by statistical agencies over time, which reflects its realized importance.

the CPI had an upward bias of 1.1 per cent per year. Of the total bias, 0.6 per cent could be attributed to new products and quality change, 0.4 per cent to consumer substitution, and 0.1 per cent to outlet substitution. In themselves these are interesting estimates. Basically they suggest that, because of new products and quality change, consumers in the US economy are becoming 0.6 per cent better off each and every year. This might be compared with the 0.8 percentage points per year calculated by Broda and Weinstein (2010) as the benefit of the introduction of new goods and replacement of the old.

In generating its estimates, the Boskin Commission did not pursue the objective of constructing a true COLI, no attempts being made to estimate the consumer surplus attributable to the introduction of new products (see chapter 12 for some discussion of such welfare issues). However, this was an issue recommended to the Bureau of Labour Statistics (BLS), together with other suggestions that dealt with the revision of its sampling procedures. Most criticism of the Boskin Commission Report, however, focused on the estimate of the CPI bias of 1.1 per cent. It should be noted that, although the 1.1 per cent estimate was headlined, the Commission suggested in fact a plausible range between 0.80 per cent and 1.60 per cent, which is in fact quite a wide range. Triplett (2006), however, criticizes these numbers as 'guesstimates'. The report stimulated further debate related to a number of involved economic, statistical, and political issues, but these are beyond the scope of our analysis here. For our purposes, the report suffices to highlight the relevance of the impact of product innovation on price index measurement and how this issue might be tackled.

The issue of the quality adjustment of price indexes has also been addressed at the EU level, especially with attempts to harmonize measurement of the CPI, as reflected in the Harmonised Index of Consumer Prices (HICP) within the EU; and it is generally accepted that such adjustment is crucial to producing a feasible and reliable measure of inflation in all European countries (Ahnert and Kenny, 2004). In the EU the methodology adopted for the HICP implies that quality adjustments should be based on explicit measurements, and only if such estimates are not available should the direct price comparison method be used.

According to the Eurostat definition (Eurostat, 2001), quality change occurs whenever the change in specification brings about a difference in utility (or functionality) that is significant to the consumer. Thus quality adjustments for, say, a new camera included in a CPI should not be based on the estimated change in production costs, but on the estimated value that a consumer derives from changes in the number of pixels or other new features of the new product. According to this definition, the suggested methodology would imply the use of explicit estimates of quality change in calculating price indexes, although how the latter are to be derived is not clearly defined. In cases where this is not possible, it is suggested that so-called qualitative equivalence is assumed. In other words, a direct comparison between the prices of different goods will be needed.

This approach contrasts with the treatment of quality change in the measurements of producer price indexes, where adjustments are made using the resource cost of the quality change. This approach follows theoretical suggestions by Triplett (1982), reinforced by Diewert (2002), who proposed the application of such a 'user value approach' on both sides of transactions (expenditure and production), thus avoiding possible mismatches in the aggregate measure of GDP.

Current practices in Europe consider both implicit and explicit methods for quality adjustments. Among the implicit methods we find direct price comparison, link-to-show-no-price-change, simple overlap, and bridged overlap. The direct price comparison of new and old items over two adjacent periods is based on the assumption that quality changes are insignificant. When a link between old and new items is in fact made in order to take quality variations into account, it is assumed that price changes reflect quality differences (link-to-show-no-price-change), although use of this method is limited to justified cases.

Overlap methods encompass both simple overlap and bridged overlap. Under the first variant, old and new models are linked in the overlap period. However, this approach can only be applied if the two products are available in the market simultaneously. This approach can be described using the following example. Consider product models *A* and *B*, where the latter is replacing the former. Crucially, it is assumed that one knows the price of both products at time t−1, that is, P_A (t−1) and P_B (t−1). The ratio of these two prices is then defined as R(t−1) = P_B (t−1)/P_A (t−1). Applying this ratio to the base-year price (i.e. to year zero price) of model *A* enables one to estimate the base-year (year zero) price of model *B*:[2]

$$P_B(0) = P_A(0) \cdot R(t-1)$$

The elementary price index at time *t* of model *B*, which has substituted for model *A*, is then calculated as

$$I_B(t) = P_B(t)/P_B(0) \cdot 100$$

indicating how the price of a product of constant quality (product B) would have been changing over time.

If the assumption of the two products being available in the market simultaneously is not acceptable, that is, when model *B* is not on the market at time (t-1) and model *A* does not exist at time t, the bridged overlap method is used. The bridged overlap method derives the quality-adjusted difference in prices

[2] Examples of such an approach can be found in Mostacci (2008).

between old and new items by looking at other items in the same sub-index (bridge items), which are available in the market simultaneously. In other words, if new products completely replace old ones, price changes may be derived from bridge items, such changes being assumed to be similar to those of the targeted products.

Whenever an original product—that is, a substantially new good without any clear-cut relationship with preceding goods—enters the market, replacements and resampling are needed if a valid index is to be produced. Mobile phones and VCRs are clear examples of such changes. In fact regular sampling revision is needed to ensure that the basket of goods included in, say, the CPI does in fact reflect changes in expenditure patterns, as changes in the goods on the market affect purchasing decisions (scanner data may facilitate the use of overlap methods and resampling procedures that allow for a better consideration of quality change; see Broda and Weinstein, 2010).

Regular replacement rates are determined by the various statistical agencies, together with so-called implicit quality indexes (IQIs) that show the adjustment needed in price indexes. Such indexes may be described by first defining a standard reference index (SRI) thus:

$$\mathrm{SRI} = \left(P_B(t)/P_A(0)\right) \cdot R_Q \cdot 100$$

where $R_Q = Q_A/Q_B$ is the ratio between the quantity of items A and B purchased and SRI is the measured raw price index reflecting relative unadjusted product prices. Thus the SRI measures price variation arising from both inflation per se and quality changes.

The ratio between the SRI and the CPI provides the IQI:

$$\mathrm{IQI} = (\mathrm{SRI}/\mathrm{CPI}) \cdot 100$$

and, because the SRI measures the price variation that depends both on inflation itself and on quality changes and the CPI (should) vary according only to pure inflationary price variations, the ratio between the two indexes—that is, IQI—provides an implicit quality measure, in that it signals whether replacements generate a significant difference in price measurements. A ratio that is greater than 1 suggests that replacements do provide higher quality.

11.4 Hedonic methods

In addition to implicit methods for quality adjustment we have also stated that there are explicit methods. These include hedonic regression and the use of option costs. We discuss the latter first, before concentrating on hedonic methods.

An example of the option cost approach is exemplified (as suggested by Camus, 2016) by considering a case in which the difference between two products may be attributable to one specific extra option. The case of cars is illustrative. Suppose a car model *A* has a different price in year 1 and in year 2, and that this difference is attributable to a new parking sensor. Given the price of the model in year 1 and in year 2 and the price of the device, one can separate the change in a car's price due to inflation from the change attributable to quality improvement.

Suppose, for example, that the price of the car is 20,000€ in year 1 and 24,000€ in year 2. Also, suppose that the price of the parking device is 1,500€. Given such information, an inflationary price change of 2,500€ may be ascribed (24,000–20,000–1,500). However, it is worth noting that this technique is only applicable when options are fully separable from the product, for only then is it possible to determine a market price for them. Also, it is commonly argued that stand-alone prices or after-market prices may be considerably higher than the implicit price of the option when attached to the new product.

The rationale behind the hedonic methodology lies in the well-known characteristics approach to consumer behaviour proposed by Lancaster (1966). In this approach, goods may be thought of as bundles of characteristics and one should be able to estimate the implicit price for each characteristic by estimating econometrically appropriate price equations. Empirically, the methodology involves the estimation of a price regression that includes product characteristics as explanatory variables. In general, hedonic regressions imply a set of estimates that relate items' prices to their measurable characteristics. In particular, they are quite widely used for personal computers (PCs), laptops, digital cameras, and other electronic goods.

To exemplify the use of this technique, consider an illustrative case taken from Wells and Restieaux (2014) that provides a clear and simple specification of the implementation of the technique to PC prices (PCs being a particularly good example). Examples of measurable characteristics for PCs include the speed of the processor, the hard disk's size, and the amount of memory. In the example, a situation is considered where a particular individual PC was priced in a reference month but is not available in the next month and in fact has to be replaced in the basket with another, similar product, but one that shows a change in specification, namely (and only) an increase in processor speed.

A regression of the log of PC prices on characteristics (the hedonic regression) across the models available on the market in some base year provides a valuation of each PC characteristic (processor speed, hard disk size, amount of memory, inclusion of monitor and video card), allowing the calculation of the PC's predicted price included in the reference month (January) as £480.87, which differs from its actual price at that time (£475.00). In the next month (February), the replacement PC is included in the bundle and is assumed to be exactly the same as the previous PC but has a faster processor speed and a higher actual price (£550). With the calculated hedonic valuations one may then calculate what the price of a PC with the characteristics of the replacement would have been (the predicted price) in the previous period—which in this case is calculated as £569.35.

Given that the hedonic coefficients are being assumed to be constant over time, one may then argue that, in January, the price of the higher quality PC (introduced in February) would have been equal to the predicted price of £569.35, by comparison to the predicted price of the actual product included in January, namely £480.87. Then the improvement in quality of the product between January and February is reflected in a price ratio relative to the older product (predicted price of the new PC/predicted price old PC): £569.35 £/£480.87 = 1.184. One may then proceed to the adjustment of the base price as the base price of the old PC × quality change: £475.00 × 1.184 = £562.40. Then, by comparing the current price (£550.00) with the new base price (£562.40), one may derive the following index of the change in price between January and February (January = 100): unadjusted PCI = actual price of new/actual price of old = 550.00/475.00 × 100 = 115.8; and adjusted PCI = actual price of new/adjusted base price of old = £550.00/£562.40 × 100 = 97.8. This simple calculation suggests therefore that, by taking into account quality adjustments, the price index has fallen by 2.2 per cent, whereas it would have increased by almost 16 per cent if such an adjustment had not been made.

This hedonic type of methodology has been partially adopted in at least ten EU countries and by the BLS in the United States; however, application of the methodology is only suitable for a limited number of goods, mainly electronic consumer durables, cars, garments, and bestselling books. In the United States the BLS has introduced a hedonic approach for measuring price movements in personal computers (an approach already in use in the construction of the producer price index since 1990). Prior to its introduction, until 1998, the weight on computers in the price index was based on consumer expenditures during 1982–4. Although in 1998 a revised weight and a new category of product were introduced, the relative importance of such goods changed little and was small (0.23 per cent). An hedonic approach was also adopted for television in 1999. However, because of matching problems, this implementation was not completely successful. Nevertheless, in 2000 new hedonic models were put in place for a variety of products, given the ability in previous periods to collect special samples of prices and product characteristics. Examples included consumer durables such as washing machines, refrigerators, and VCRs. Johnson et al. (2006) show that the impact of such approaches has been relatively small and the impact of quality adjustment did not differ dramatically from that of standard implicit adjustments.

Cellular phones provide another significant example of the need to incorporate quality and price adjustments into CPIs. Hausman (1999) estimates that the omission of such products in the CPI until 1998 (when they were finally included) brought about an upward bias in the telecommunications service price index of about 0.8 per cent to 1.9 per cent per year. More precisely, Hausman (1999) calculates that, instead of a 1.1 per cent increase in the telecommunications service price index, one should have observed a 0.8 per cent decrease per year, which caused a significant bias in the calculation of the overall price index.

Following the outcome of the Boskin Commission, Eurostat also set up a task force designed to look at price deflators that, in particular, underlined the large differences in information and communication technology (ICT) deflators used across European countries (Eurostat, 1999). This issue is relevant to both the harmonized measurement of inflation and the growth rate of productivity, which would be subject to possible measurement errors if different deflators for ICT investment[3] were used. The EU report concludes that the differences were mainly dependent on differences in measuring quality improvements, thus implicitly contributing to the introduction of hedonic price estimation in Europe.

In the United Kingdom the use of hedonic techniques is limited to electronic products, for example PCs, laptops, tablet PCs, digital cameras, smartphones, and mobile phones; these techniques have been introduced gradually into calculations of the Retail Price Index (RPI). Such items with high replacement rates (the replacement ratio is the ratio of the number of replacements to the sample size in the calculation of a price index) and significant quality changes bring about significant price and quality adjustment issues, and therefore some such correction is especially important for them. PCs and laptops fall within this category of goods, showing high levels of quality change and replacement. In United Kingdom's CPI, the replacement ratio was 3.1 and 2.9 respectively for PCs and laptops in 2013. However, one might note that another relevant issue is represented by the weight of new products in the overall basket of goods; and the sample weight of electronic products like cameras and mobile phones has fallen over the last ten years. In the United States, in addition to electronic products, hedonic techniques are applied to clothing, footwear, refrigerators, washing machines, microwaves, clothes dryers, ranges, and cook tops. Sweden and Germany have also introduced the hedonic methodology, which encompasses (in addition to electronics) footwear and clothing in Sweden and used cars in Germany.

Hausman (2003, 1999) criticized the hedonic approach on the grounds that it is unable to provide a true COLI. This consideration is relevant, as it raises the issue of how closely the measured consumer price relates to the true cost

[3] Van Ark, Inklaar, andMcGuckin (2003) present estimates of ICT investments and growth accounting that are based on quality-adjusted ICT deflators.

of living.[4] A typical hedonic regression shows a product price as the dependent variable and the product's characteristics as covariates. Once the coefficients are estimated, it is possible to adjust the observed prices for changes in the product's characteristics. As an example, Hausman (2003) suggests that such price adjustments are not related to the consumer's valuation of a computer. In other words, hedonic regressions do not represent structural economic equations, in that they do not capture the mix of factors that affect prices, for example mark-ups over costs and factor input prices, which in fact are mainly related to firms and consumers' characteristics and are also dependent on time. For example, (Berndt et al., 1995) find that coefficients derived from hedonic regressions of PC prices are crucially affected by time and, interestingly, suggest that it might be better to also consider prices on used computers in order to model their price dynamics in response to technological changes embodied in new models.

Overall, although controversial, hedonic regressions represent a practical and relatively easy method of explicitly taking into account quality change in a product environment that is characterized by quality change and high volatility.

11.5 Conclusions

We have analysed how quality change may affect price measurement. The accuracy of such measurement is important because it affects the accuracy of the calculation of both inflation and real variables such as the GDP growth rate. The work of the Boskin Commission stimulated debate on quality adjustments at the international level. Technological innovation and, in particular, the impact of ICT technologies underline the relevance of this issue, as does the importance of international coordination in the production of reliable and updated measures of price inflation and real variable growth.

Two main changes over time related to product innovation are important in this exercise. The first is that new goods (which are often cheaper) are driving old goods out of the market, and, secondly, new products often offer improved quality. Broda and Weinstein (2010) suggest that a failure to properly account for these changes has added 0.8 percentage points per year to the measured CPI in the United States, whereas according to the Boskin Commission it might have added 0.6 per cent per year. Another way of looking at this is that the direct benefit to consumers in the US economy from the introduction of new goods of higher quality and lower prices as a result of product innovation may equal at least 0.6–0.8 per cent per annum.

[4] Deaton (1998) argues that the consumer value of new goods should not be included in a CPI.

Quality adjustment approaches in Europe and the United States in all OECD countries have converged towards general methodological guidelines that represent a common knowledge base. The hedonic methodology is being applied in a significant number of countries and to specific categories of goods, in particular electronic products. We have exemplified the use of this approach and evaluated its impact on price indexes. It is clear, however, that more work is required if unbiased measures of quality-adjusted price changes over time are to be widely available. In particular, further coordination in Europe is needed in the harmonization of price indexes.

12 Product Innovation and Welfare

12.1 Introduction

In chapters 10 and 11 we have addressed how product innovation may impact upon prices and firm performance, including profits, output, sales, and productivity. Clearly all these factors will affect economic well being, that is, economic welfare. The aim of this chapter is to address welfare issues more directly. This is done via two approaches. The first is to explore the impact of product innovation upon a commonly used measure of economic welfare, calculated as the sum of consumer and producer surpluses. The second approach is to explore how product innovation affects economic welfare via its impact on the degree of product variety in the economy. Throughout the chapter we attempt to address two main issues. The first is the impact of product innovation upon welfare; the second is the question whether a free market economy, unaided, not only can produce improvements in welfare but also has the necessary incentives to generate welfare optimal outcomes. The results of this discussion will provide some foundation for the discussion of policy intervention in chapter 13.

12.2 Product innovation and consumer and producer surplus

12.2.1 *Repeat purchase products*

When a new product is introduced, it may either create a completely new market (the cross-price elasticity of demand with existing products is very small) or enter an established market (and have a higher cross-price elasticity with existing products). The new product may also be durable or non-durable (and thus a possible repeat purchase). We first consider the simplest case, where a completely new, non-durable, possibly repeatedly purchased product is introduced to the market with no past or current competing substitutes. We explore the durable good (non-repeat purchase) case in section 12.2.2.

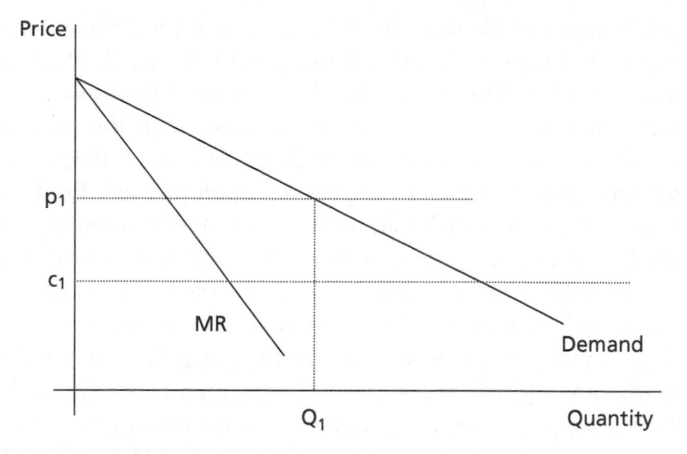

Figure 12.1 Product innovation, consumer and producer surplus

In Figure 12.1 we plot the demand curve for a completely new, non-durable, product and the associated marginal revenue curve. Let the marginal and average cost of production of the product be c_1 and let the price charged be p_1. At price p_1 let the number of units demanded and sold be Q_1.

Consumer surplus (CS) is measured as the area under the demand curve at (p_1,Q_1) and represents the sum of the amounts that buyers would have paid for the product over and above p_1; thus it gives a measure of their valuation of the product over and above the cost of purchase to them.[1] Producer surplus (Π) is a measure of the revenue earned by suppliers minus the cost of production, that is, profits, and is measured by $p_1Q_1 - c_1Q_1$. The total welfare arising from the introduction of the new product is thus equal to CS plus Π.

Simple manipulation of the curves in Figure 12.1 yields two obvious results. *Ceteris paribus* (in particular for a given price), the further the demand curve lies to the right, the greater will the welfare gain (both CS and Π) arising from the new product be—that is, the more the new product is desired by potential buyers, the greater is the economic benefits that it yields. Secondly (again, all other factors being equal), the lower the costs of production, the greater will both the consumer and the producer surplus be, and thus the greater the welfare gain from the new product. Less obvious is what will happen as the slope of the demand curve changes. As long as the vertical intercept is not lowered, a flatter demand curve will also lead to a greater welfare gain (given costs and price) but if the intercept is also reduced then the net result is not clear.

[1] It can be noted that the buyers may be producers rather consumers, in which case this surplus would be equal to the profit gain made from having the new product.

What will happen to welfare, given the costs and the demand curve, if the price charged is changed? Clearly, if the price is reduced, then Q_1 will be greater, and thus CS will be greater too. In fact CS will continue to increase as the price falls towards c_1 and will be maximized when the price equals c_1. However, reductions in p_1 and the resultant increases in Q_1 will impact differently upon producer profits, depending upon p_1 itself. Producer profits are maximized where p_1 is such that the value for Q_1 equates marginal revenue and marginal costs, c_1. If p_1 exceeds this level, then a reduction in price will increase the profits; however, as this is the case, no supplier would charge such a price to start with—the point where marginal revenue and marginal costs are equal being a monopolist's profit maximizing price. If p_1 is smaller than at the point where marginal revenue and marginal costs are equated, then, as price falls, the supplier's profits decline. Thus we may expect that, as price falls, CS increases and Π declines. The sum of CS and Π, however, being the area between the cost curve and the demand curve at Q_1, will always increase with Q_1 (if p_1 is smaller than at the point where marginal revenue and marginal costs are equated), and thus welfare arising from the new product will increase as its price falls, but with CS increasing and profits decreasing. Welfare measured as the sum of consumer and producer surplus will thus be maximized when the price of the new product equals its marginal cost of production.

A particular variation on this type of case concerns the gain arising when the new product partially or completely supersedes an existing product (or products). Allow that prior to the product's innovation the market was served by a product selling an amount Q_0 at price p_0 produced at costs c_0, and generating consumer surplus CS_0 and profits Π_0. If the new product completely supersedes the existing product, then the extra welfare from the innovation is equal only to the gain in welfare and not to the whole welfare generated by the new product, that is, to $(CS + \Pi) - (CS_0 + \Pi_0)$. One might note, however, that this is a flow calculation and not a stock calculation. There is the possibility that the replacement of the old by the new involves losses in the value of human and physical capital, and perhaps knowledge capital that may not be accurately reflected in the reduction in producer surplus. An example may be how mobile phones and computer-based communication such as e-mail have decimated the demand for standard telephone landlines and postage systems and thus reduced the need for sorting offices, existing mail distribution, telephone systems, and postal workers.

Clearly, if the new product only partially replaces the existing product(s), then the effect is not so severe, but it is still only the gain in welfare that should be counted as arising from the new product. The nature of the outcome under partial replacement depends to some degree on whether the existing product is horizontally differentiated from the old or vertically differentiated from the old. In the latter case we tend to think of a market where the buyers differ not according to their taste but in terms of their ability to pay. If the new product is of superior quality, previous purchasers of the existing (and now lower quality) product who have the ability to pay will shift their demand to the new, higher quality product, and thus the launch of the new will shift the demand curve for the existing product to the left (the extent of the shift depending upon the price charged for the new product). If the shift of the demand curve for the old product is such that, when the new product is sold at the monopoly maximizing price (i.e. where MR = MC for the new product), the demand curve for the old product cuts the vertical axis below its marginal/average cost of production, it can no longer be produced at a price at which it can sell, and so it will be removed from the market. However, if the improvement of the new product over the old is not so great, or if the ability to pay for higher quality product is low, then the old product may remain on the market. But the demand curve for that product will shift to the left (partly depending on the price of the new product) and both its selling price and the quantity sold will decline. The launch of the new product will then lead to a market with a greater menu of quality types than previously.

In the case where the new product is vertically differentiated from the existing product but is of lower quality, for it to succeed in the market it must be sold at a lower price than that of the existing product. It must thus essentially enjoy lower costs of production. Its appearance will, again, shift the demand curve for the existing product to the left. This may lead to the exit of the existing product from the market but might just generate a situation where there is a wider spread of quality offerings on the market. Once again, however, the welfare gains from the innovation should be measured as net of losses to welfare arising from existing products.

In the case of horizontal differentiation, we conceptualize a market where buyers have different preferences over the characteristics of products. The new product is considered to offer an alternative combination of characteristics to those already on the market. In this case, once again, there is the chance that the new will not completely displace the old. This will depend on the degree of overlap between markets—for the existing product and for the new product respectively—and on the costs of manufacturing the new and the old. Thus, although demand for the existing product may well be reduced by the appearance of the new, the costs of producing the new may not be sufficiently low to enable a price to be charged that would encourage those with preferences more distant from the character of the new to swap from the existing product in such large numbers as to make it non-viable.

One reason why the impact of the new product on the continued existence of the old matters is that the longer the existing products remain in the market the greater is the competition that the innovator faces. This is important, for the degree of innovation in the market can have a considerable effect on the welfare gains generated by the product innovation. In particular, one may argue that competition will limit the extent to which the price of the innovative product can exceed marginal costs. In terms of Figure 12.1, one may expect that p_1 will decline as the number of competitors in the market increases. As this occurs, Q_1 will increase. As already argued, as price falls, CS increases and Π declines for the product innovation. The sum, however, will always increase, and thus welfare arising from the new product will increase as its price falls. Competition thus increases the welfare gains arising from the new product. Using a similar argument, if there is emulation of the innovators by others, then, again, price will fall, causing CS to increase and Π to decline. In addition, the copies launched on the market will further increase welfare through their own generation of profits and CS.

There have been attempts in the literature to quantify the sort of welfare gains that arise from product innovation (as discussed here). Not surprisingly, this is no simple task. There is a large literature exploring how innovation may impact upon profits (see chapter 10), but very little on how it impacts upon CS. It is clear that what is required in order to calculate the impact of product innovation on the sum of consumer and producer surplus (which are in fact jointly determined) is some estimate of the demand curves, of the cost curves (or production function), of the degree of competition in the market, and of pricing. Early attempts at this exercise are Trajtenberg (1989), who explored both the static and dynamic welfare gains from computed topography scanners, and Hausman (1996), who looks at new cereal brands in the US market. Using sophisticated econometric methods, they estimate demand relationship and other factors and show that considerable increases in welfare are generated. More recently, Jaumandreu and Mairesse (2010) develop a simpler framework for estimating the parameters of the production function together with the elasticity of demand for the output and the impact of demand and cost shifters. Using data on about a thousand Spanish firms in six industries between 1990 and 1999, they estimate that the total current period (static) welfare gains of introducing a product innovation are about 4 per cent of the value of the firm's current sales, this value being split almost equally between consumer and producer surplus.[2] (For comparison, process innovation is estimated to increase welfare by 1.6 per cent, two thirds being captured as CS.)

Such estimates as these—and, it is fair to say, as the analysis presented here—do not, however, take explicit account of the open nature of most economies with goods that are both imported and exported. If the new product is imported rather than produced at home, then the producer surplus will be received by the overseas producers rather than by the domestic producers. In a national rather than a global sense, then, welfare gains from the product innovation will be equal only to the CS realized. This will clearly

[2] Jaumandreu and Mairesse (2016) illustrate the difficulties of this exercise, particularly of separating out of the impact of process and product innovations on firms' cost and demand functions.

increase as the selling price falls or as the demand curve moves to the right, without any offset as a result of a falling producer surplus. If, on the other hand, the product is produced at home but sold overseas, then the CS does not yield any domestic benefit. Reductions in price will thus yield benefits overseas, but the reducing producer surplus will cut welfare domestically. Añón Higón and Stoneman (2011) make some attempt to measure for different countries how international trade affects the calculation of gains from innovation and they show that, although it varies by country, for some countries up to 20 per cent of any gains may derive from imports and a similar amount be lost via exports.

Putting to one side this issue of exports and imports, one may proceed to address not only whether product innovation produces a welfare gain and when that gain is maximized but also whether that maximum will be realized in a free market. From Figure 12.1 and the ensuing discussion it is clear that the sum of consumer and producer surplus will be greatest when the price of the innovative product equals its marginal (equals average) cost of production ($p_1 = c_1$). Will a free market generate this outcome unaided? One may argue that, as competition in a market increases, the price charged in that market will fall, price will tend towards marginal costs, and the sum of consumer and producer surplus will tend towards the maximum. In a market with completely free entry, where an innovative product may be quickly emulated and copied by others, this may be the expected outcome. However, if there are restrictions in the market that limit entry, for example secret knowledge, appropriated skills, and so on, then there may not be quick emulation and the market outcome as a result of profit seeking by producers may involve a price for the innovative product above marginal cost and a sum of consumer and producer surplus below the possible maximum.

12.2.2 *Durable products*

Many new products are durable, having a possibly useful life of many years. In such cases discussions of welfare must encompass the intertemporal dimension of their production and distribution, in addition to the issues already discussed. To illustrate, consider a product innovation of which only one unit will be purchased by any buyer and that has an infinite physical life, yielding a typical purchaser a gross benefit from ownership of g per period from the date of purchase. Define year zero as the date the product is launched, and year t as the date of its acquisition by the typical purchaser at price p(t). The gross lifetime discounted benefit of ownership to a typical purchaser evaluated at time t will be equal to g/r, where r is the discount rate and the benefit net of acquisition cost will be $g/r - p(t)$. Evaluated at time 0, this benefit will have a discounted value of $(g/r - p(t))/(1 + r)^t$, which is the CS from acquisition by the typical purchaser evaluated at time 0.

The production of the innovation in time t is assumed to incur costs of c(t). If the product sells at price p(t), the profit of the innovator from selling to the typical consumer will be p(t) – c(t), which has a present value in time 0 of $(p(t) - c(t))/(1 + r)^t$. The welfare thus generated by the production and sale of the innovation to the typical consumer, evaluated at the date of launch of the innovation and defined as w, will equal the sum of consumer and producer surplus, that is $(g/r - p(t))/(1 + r)^t + (p(t) - c(t))/(1 + r)^t = (g/r - c(t))/(1 + r)^t$. Summing w over all acquirers will yield the total welfare gain (consumer and producer surplus) from the new product.

Let us assume initially that there is population size N of potential acquirers and all such adopters face a common per-period benefit of g. Let us also assume that c(t) and p(t) stay constant over time, at c and p respectively. Assuming that $g/r - p > 0$, all consumers will wish to buy the new technology; and assuming that $p - c > 0$, producers will wish to supply all consumers. With constant prices, that is, costs of acquisition, and identical acquirers, all N consumers, if appropriately informed of the innovation, will wish to acquire at a common date, the most beneficial one being when g/r first exceeds p, which is in fact the date of launch of the new product. In such a case the total welfare gain from the product innovation evaluated at time 0, W, will equal $N(g/r - c)$. Obviously, *ceteris paribus*, the greater the size of the population and the benefits gained from the innovation, g, and the lower the costs of production, c, the greater will the welfare gain be.

However, we know that, by and large, durable new products experience a diffusion process, in that not all purchasers buy at the same time and acquisition is spread over time. One way to rationalize a spread of acquisition over time if prices and costs of acquisition are not changing and g is the same for all potential acquirers is, for example, to hypothesize some information deficiencies in the market such that some potential acquirers do not hear of the innovation until after others. In such circumstances (assuming that $g/r - p > 0$), buyers will acquire the new technology at the first date when they are informed of its existence. Let the number of buyers informed at time t be written as x(t); then x(t) is also the stock of the new product acquired at time. The number of those who acquire the product at time t will then equal dx/dt, which we write as q(t), each buyer benefitting from a discounted benefit of ownership evaluated at the launch date of $(g/r - p)/(1 + r)^t$ and supplied at a cost of c. Thus, for any t, the welfare generated equals $(g/r - p) - c)/(1 + r)^{t\cdot}\, q(t)$ and the total welfare, W, arising from the product innovation equals the sum of this value over all t. We may now observe (in addition to the effects of a greater population, larger g, and lower c, on the welfare gain) that, as q(t) changes, the welfare gain will change too. In particular, if there are more early adopters and fewer later adopters (perhaps because of information spreading more quickly), then the population will obtain the benefits of the new technology earlier and welfare (the sum of consumer and producer surplus) will be increased. Of course, the earlier adopters will also be gaining more than the later adopters.

In such a story, however, the *deus ex machina*—the exogenous technique for spreading acquisition over time (information deficiencies)—can be improved upon. In doing so we can gain more insight into the intertemporal acquisition patterns and welfare benefits from product innovation. To do this, however, requires considerably more structure. We allow that potential acquirers differ from one another, such that some gain more benefit than others. Assuming full information, we may take as given that those who obtain greater benefit will acquire the product before those who obtain smaller benefit. Defining $x(t)$ as the total number of those who acquired the new product by time t, we write that the per-period benefit obtained by the xth acquirer is $g(x)$, assumed to be constant per period[3] once the product is acquired and to decline with x at the date of acquisition. The gross benefit of acquisition, evaluated at time t by an acquirer who acquired at that time, will have a discounted value of $g(x(t))/r$ and acquisition will depend upon the difference between this and the cost of adoption $p(t)$ in time t. It is reasonable to assume that a potential acquirer will adopt the innovation if the benefit from doing so is greater than the cost,[4] that is, if $g(x(t))/r - p(t) \geq 0$. The marginal adopter at time t will be that buyer for whom $g(x(t))/r = p(t)$. It is immediately clear that (i) if the number of adopters is going to increase over time, then the cost of adoption $p(t)$ must be falling over time; (ii) the gross benefit obtained by the next adopter declines as the number of adopters increases; and (iii) if more early adoption is to occur, then it must involve lower costs of adoption for potential adopters and thus marginal adopters who obtain smaller gross benefits. However, as each user generates benefits greater than (or equal to) the cost of adoption, any increase in usage brought about by reductions in this cost must generate an increase in CS.

If increasing usage over time requires that the cost of adoption declines, the margin earned by a producer facing costs that are invariant over time must be declining. Speeding up adoption (bringing it forward in time) will require that prices are reduced more quickly, in at least some time periods, and in consequence the margin per unit sold will be reduced further. That is, faster adoption will require a reduction in producer surplus. The limit to this process is when the price of the innovation equals its cost of production, at which point extensions of usage end. Thus bringing usage forward in time will increase CS but decrease producer surplus. However, as $p(t)$ is greater than the cost of production c, producer surplus will be positive for each unit sold

[3] A variant of the model not explored here could allow the benefit per period to decline with the stock of users during that period.

[4] Assuming that there is no intertemporal arbitrage, i.e. the buyer does not take into account potential future price reductions, sometimes labelled myopia.

and earlier sales will increase welfare, but each extra sale induced will add less to welfare than the previous unit sold.

A further extension of this discussion of the durable product case considers a situation where the new product is itself improving over time. This might be through the launch of improved product variants embodied in further innovations. In such a case the benefits obtained by a purchaser in time t will be dependent not only upon that adopter's characteristics (and hence on his/her position in the order of adoption) but also upon his/her date of adoption. It is interesting that Trajtenberg (1989) looks specifically at the social gains arising over time from the adoption of an improving technology (computer topography) and calculates that the social gains generated by the product innovation in the first half of the period studied are far larger than those generated in the second half. The profile of gains over time can be represented as a sort of lognormal distribution that starts high, rises still further during the initial period, and then declines rapidly, carrying forward a low-level tail.

In such a model, once again, the adoption dates could be brought forward in time, perhaps through reductions in the cost of acquisition, which, as before, may increase welfare if the price of the innovation is still above the costs of production. However, there is a downside in that the technology being acquired via earlier adoption will be inferior to the one that would have been purchased with later adoption. The welfare gain will thus be partly offset.

The final intertemporal scenario that we consider is one where the product itself is not improving but the cost of supplying the product $c(t)$ is declining over time. In essence, as time proceeds, the cost of production declines, so that the price to be paid by potential adopters falls, and this leads to extension in ownership. In such a model, if adoption is brought forward in time by a price reduction at time t, then this would increase the CS (assuming that for the marginal adopter $g(x(t))/r \geq p(t)$); however, the producer surplus would be lower, both because the price would be lower and because the extra unit would now be supplied at a higher cost than it would have been otherwise.

This particular intertemporal framework has been explored by Ireland and Stoneman (1986), first in order to characterize the nature of the intertemporal adoption pattern that maximizes the (discounted) sum of the flow of consumer and producer surpluses, and then in order to explore whether this pattern will be realized in a free market. It is shown that on the adoption path that maximizes the sum of consumer and producer surplus (W), at any time t, when $x(t)$ units have been sold, the benefit obtained by the xth adopter $g(x(t))$ will be equal to $rc(t) - dc(t)/dt$. We argue that, if buyers do not undertake intertemporal arbitrage and either ignore or are unaware of changes in the cost of adoption over time (i.e. if they are myopic), then the marginal adopter in time t will be that buyer for whom $g(x(t))/r = p(t)$ and thus, on the W maximizing path, prices over time will be given by $rp(t) = rc(t) - dc(t)/dt$. Ireland and Stoneman (1986) show that this W maximizing path would be the price path chosen by a profit-maximizing monopolist. A monopolist supplier would thus produce the W maximizing diffusion path. In addition, in such a scenario it is shown that, on this path, producer surplus would also be at a maximum.

Ireland and Stoneman (1986) also go on to show that, if potential buyers do undertake intertemporal arbitrage and thus realize and act upon the expectation that the costs of acquisition will fall over time (implying that, for the marginal buyer, $g(x(t)) = rp(t) - (dp/dt)$, where dp/dt is the expected change in price), then a profit-maximizing monopolist would not generate the W maximizing path and in fact would generate an intertemporal adoption pattern that would be too slow on the basis of this criterion. In such a scenario, however, if there were a number of competing suppliers, then their attempts to capture customers before their rivals (sometimes called the common pool effect) would cause them to generate the price and adoption path that maximizes the sum of consumer and producer surplus.[5] In this scenario, however, producer surplus is zero and all the surplus is taken by the consumers. Thus, although monopoly supply and myopic demand generate the same maximum sum of consumer and producer surplus as competitive supply and non-myopic demanders, the two scenarios generate very different distributions of the surpluses between consumers and producers.

12.2.3 *The distribution between consumer and producer surplus*

We have considered welfare as measured by the sum of consumer and producer surplus and proceeded to explore how product innovation will impact upon this measure and when it might be maximized. We have also pointed out that, under different scenarios, the distribution of the total between consumer and producer surpluses will be different. In some cases increases in the total may be associated with greater producer (consumer) surplus, in some cases with smaller surplus. It is reasonable to argue, however, that the distribution is as important as the total and as a result one cannot necessarily always consider that the welfare optimum is characterized by that position that maximizes the sum of consumer and producer surplus. To illustrate, in the simple repeat purchase scenario, the sum of consumer and producer surplus is maximized when the innovative product is priced at its marginal cost. At this point producer surplus (profit) was zero. However, zero profits will not provide an incentive to the innovator to develop and launch

[5] It may also be shown that, if buyers are myopic, then competition in supply will generate an adoption path that is too fast and the sum of consumer and producer surplus is not maximized.

new products for the market[6] (see Arrow, 1962). Unless there is such an incentive, the new products will not be developed and the welfare gains that they would allow will not be realized. Thus the welfare optimal position will not necessarily be the one that maximizes the sum of consumer and producer surplus. In the durable product case we showed that the sum of consumer and producer surplus was maximized either when buyers were myopic and there was a monopoly supplier or when buyers undertook intertemporal arbitrage and there was competitive supply. In the former case, however, supplier profits were maximized and in the latter case they were minimized. The two scenarios would thus provide different incentives for developing product innovations.

Economics generally argues that a market will generate a welfare optimal outcome when the incentives for an action equal the social benefits that will arise from that action. It has long been realized that the social benefit from innovation will in most cases exceed the private benefit from an innovator's action. Unless the innovator is able to perfectly price-discriminate (as in certain scenarios in the intertemporal story considered here), even if there is a monopoly supplier who is able to maximize his/her own profits, there will still be a CS arising from innovative activity, and as a result the social return (the sum of consumer and producer surplus) will exceed the private return (producer surplus) and the incentives to innovate will be lower than the social benefit. This is suboptimal. This situation would be further exacerbated if the innovator is copied by others, thereby reducing his private return while increasing the social return, or if there are spillovers and externalities arising from an innovator's activity that yield returns to society and not to the innovator. It is thus frequently argued that the private return to innovation is lower than the social return, and as a result investment in innovative activity will be suboptimal (for a review of estimates of private and social returns to R&D, see Hall et al., 2010).

Although the argument just presented would seem to suggest that, as producer surpluses increase, the market failure will reduce and welfare would improve, this is not necessarily so. A particular argument arises when innovation is a race and the first to market takes all. This can be the case when there are inviolable patents. Such situations may encourage races in the development of new products that involve over-costly speeds of development (Dasgupta and Stiglitz, 1980), because the players are concerned with who wins the race, whereas society is not (Dasgupta and Stoneman, 1987). Very large returns may also encourage excessive innovation, generating (i) suboptimal patterns of early product replacement, (ii) repetition of technological development, or (iii) excessive product variety. Thus, although zero producer

[6] The incentive for the innovator will equal the difference between profits with innovation and profits without innovation (thus allowing for the possibility that others will innovate), minus the cost of developing the innovation.

surplus is unlikely to be optimal, maximized producer surplus might not be optimal either. Something between the two may be preferable.

12.3 Product variety

In addition to product innovation generating increases in consumer and producer surplus and thus in welfare, product innovation will also impact upon product variety, that is, upon the extent of choice of buyers among products available on the market. Increases in buyer choice will have a positive impact upon welfare. The increases will come about through the availability of either more choice per se, or products more suited to personal taste or more affordable given buyers' incomes. There are, however, costs related to this increase in variety: (i) the production of most products will involve some fixed costs, and the greater the number of products on the market the greater the resources required to cover fixed costs; (ii) the development of new products will also involve costs, and the more products the greater the resource costs of development; and (iii) if there are increasing returns in production, more variety may mean lower sales for any particular variety and perhaps higher individual production costs. In looking for the overall benefit of variety, one has to compare the costs and the gains.

In chapter 4 we have discussed how industries change over time and how, with the introduction of new products and the removal of old, the number of products on the market changes as markets age and mature. In chapter 7 we explored models of horizontal and vertical product innovation, in an attempt to address the positive economics of this process. The analysis of such models suggests that, at any time, the total number of product variants on the market, the number of new product variants that are launched on a market, the locations in product space, the prices, the quality, and the expenditure on new product development will reflect the nature of competition, fixed costs, demand elasticities, the distribution of consumer preferences, the costs of development, the market structure, and many other factors. The different scenarios make general predictions rather difficult, except that it is expected that expenditure on product innovation will continue until the extent of product variety reaches the point where the marginal gross payoff of introducing a further new product equals its marginal cost of production. At this point an existing product will be removed from the market as its marginal costs of production can no longer be covered by its selling price. This point, however, is determined by many factors.

We do not have any estimates of the contribution that the extent of variety and changes in variety make to welfare. Instead, we may ask a related but slightly different question. Will the market, unaided, create a welfare optimal extent of product variety? Lancaster's (1990) survey of the literature on this matter provides a (still) surprisingly definitive view of the difficulty of answering this question:

> The fundamental structure of all optimal variety problems, for the individual firm as well as society, is the interplay of two elements in the economy-the existence of a gain from variety and the existence of scale economies of some kind. If there are no economies of scale associated with individual product variants (in distribution as well as in production), then it is optimal to custom produce to everyone's chosen specification. If there is no gain from variety and there are scale economies, then it is clearly optimal to produce only a single variant if those economies are unlimited, or only such variety as uses scale economies to the limit (all products at minimum average cost output). Most cases involve a balance of some variety against some scale economies, the solution depending on the preference properties of consumers, the scale properties in production and distribution, and the way in which the social welfare criterion is derived from individual preferences. Different criteria and assumptions can lead to quite different conclusions. (Lancaster, 1990, p. 204)

Specifically, therefore, different models will have different normative predictions. In some scenarios expenditure on developing variety may be too low. In others it may be too high. Dixit and Stiglitz (1977) address this issue in the context of their horizontal product differentiation framework. Their key finding is that there is some presumption that the market solution would be characterized by too few firms in the monopolistically competitive sector; but their fuller findings reflect the complexity of the answer to the main question:

> In the central case of a constant elasticity utility function, the market solution was constrained Pareto optimal, regardless of the value of that elasticity (and thus the implied elasticity of the demand functions). With variable elasticities, the bias could go either way, and the direction of the bias depended not on how the elasticity of demand changed but on how the elasticity of utility changed. We suggested that there was some presumption that the market solution would be characterised by too few firms (i.e. product innovation) in the monopolistically competitive sector. With asymmetric demand and cost conditions we also observed a bias against commodities with inelastic demands and high costs. The general principle behind these results is that a market solution considers profit at the appropriate margin, while a social optimum takes into account the consumer's surplus. However, applications of this principle come to depend on details of cost and demand functions. (Dixit and Stiglitz, 1977, p. 308).

Fantino (2008) shows that, in a horizontally differentiated market framework where firms can invest in R&D (to develop new products) and can modify the level of differentiation of their products, increasing their specialization and their market power, firms will underinvest in R&D, because they do not internalize the effects of their research effort on the overall level of substitutability of the other varieties and on the profits of the other firms.

In a model of vertical product innovation, Greenstein and Ramey (1998) argue that competition and monopoly in the old product market provide identical returns to innovation when (i) the monopolist is protected from new product entry and (ii) innovation is non-drastic, in the sense that the monopolist supplies positive quantities of both old and new products. If the monopolist can be threatened with entry, monopoly provides strictly greater incentives. Reisinger (2004), however, argues that welfare is not necessarily improved when entry is freer rather than restricted. In a model of vertical product differentiation it is argued that potential entry may cause the quality of products to be lower if qualities are strategic complements; but, if firms can produce a quality range and practice non-linear pricing, welfare in the case of entry deterrence is higher than under monopoly, because the incumbent enlarges its product line. If entry is accommodated, consumer rent increases but the consequences on welfare are ambiguous. Lambertini and Orsini (2000) argue more definitively that the incentive towards both product and process innovation is always larger under social planning than under monopoly, and thus monopoly provides too little incentive to innovate. Overall, therefore, there is some support for the view that, left unaided, markets might generate insufficient variety; but the findings vary with the situation and are not unambiguous.

12.4 **Conclusions**

In this chapter we have addressed issues relating to the impact of product innovation on economic welfare. Initially we defined welfare as the sum of consumer and producer surpluses. We then looked at non-durable products that may be repeatedly purchased and showed how product innovation can increase welfare via additions to CS and larger firm profits, and we reported an estimate from the literature that the value of the increase for a typical product innovation might equal 2.5 per cent of the innovator's revenue. We pointed out, however, that there were problems with measuring welfare by the sum of consumer and producer surplus because, although reductions in market price always lead to an increase in this total, the allocation between consumers and producers would change, consumers getting more and producers getting less as the price fell. This would reduce the producers' incentives to innovate. We then looked at markets for durable goods, seeking insight into the welfare optimal allocation of purchases of a product innovation over time. Using the discounted sum of consumer and producer surpluses as the measure, we showed that the optimal path could arise under either monopoly supply or competitive supply, depending on buyers' price expectations formation processes. However, under the former producer surplus was maximized, and thus the incentive to innovate was greatest, whereas under the latter the CS was maximized, and there were no producer surpluses to stimulate innovation. Finally, we considered that variety may also generate welfare and asked whether markets would generate optimal variety. The literature suggests not.

13 Product Innovation and the Policy Dimension

13.1 Introduction

The purpose of this chapter is to explore why, how, and with what effect governments intervene in the process of product (and other) innovation. In so doing we mainly refer to UK policies, although it is worth noting that such interventions are now widespread in most developed, and often developing, economies and in consequence there are many different policies in operation in different economies, each with its own nuances and details. We do not attempt to address issues of detail but instead consider the generic aspects of different policies; hence we do not try to make international comparisons.

Product innovation can improve firm performance, lead to lower quality-adjusted prices, improve variety, and generate increased economic welfare (see chapters 10, 11, and 12). Other literatures would also suggest that innovation can improve trade performance (Cassiman et al., 2010), and also (especially product innovation) lead to higher employment (Edquist et al., 2001). In these circumstances it is no surprise that commentators and politicians argue that, if innovation can generate such benefits, then we ought to have more of it, in other words innovation is good, so more must be better. Of course this is, from an economic viewpoint, a non sequitur. Innovation may be good, but generating more requires greater resource inputs and this may not necessarily yield a net benefit. This rider is not frequently added.

An alternative, but related, stance is built upon international comparisons of innovation performance with governments of economies that are lower in international league tables, considering that it is appropriate to pursue policies to improve their national performance. But such tables are not necessarily based on the best comparators. For example, a very common indicator used in such tables relates to R&D spending. However, R&D is an indicator of input to the innovation process, not of an output, and it is the output from the process that should concern governments. Nor is it clear that the indicators employed treat all economic sectors equally: there is a tendency for such measures to be more appropriate for measurement in the manufacturing than in service sectors or in the creative industries. Having said this, however, these two approaches (more is better, and international comparisons) have often provided a foundation for much innovation policy.

The third approach to providing a grounding for policy is to argue that intervention is required when, if left unaided, the market is failing or will fail. This market failure argument has become more popular with commentators and politicians recently but has on occasion been rather subverted to particular ends. Market failure is now sometimes employed to justify intervention when, without intervention, the market appears to produce an outcome that is not to the liking of the commentator (which, as is immediately obvious, could vary according to political affiliations). In fact, market failure is a technical term with groundings in economic theory and refers to situations where, without intervention, the market will not produce the welfare optimal outcome, that is, the outcome that maximizes economic welfare. As seen in chapter 12, economic welfare in a static sense can refer to the sum of consumer and producer surplus, but, in an intertemporal view, could encompass the sum of both present and future (discounted) surpluses. It may also involve the extent of product variety generated by the market. There is a strong theme in the literature that suggests that in a free market economy the private return to the innovator will be lower than the social return and, in consequence, there will be market failure and insufficient investment in innovation. If the grounds for policy are based upon market failures, then the appropriate policy may be more clearly indicated than if the justification is just to attain more innovation.

There are many policy instruments that may be used to influence the innovation process. There are also many policies that, although not primarily targeted at innovation activity, may well impact upon that activity. Our purpose here is to concentrate on policies that are primarily directed at product innovation and, as such, rule out from discussion some of the policies with wider impact. Thus we do not discuss the relevance of the rule of law and democracy, although it may be of crucial importance (see, for example, Rigobon and Rodrik, 2005). We also exclude discussion of general fiscal and monetary policy, for, although the monetary environment (e.g. interest rates) and attempts to balance the budget (perhaps via reductions in government spending) may well impact upon the innovation environment, these policies tend to be driven by the macroeconomic environment rather than by a desire to stimulate innovation. This does not mean that we rule out consideration of more directly innovation-targeted policies (such as tax incentives to R&D), which impact on the budget surplus, only that the decisions on the size of that surplus are not considered.

We also rule out, perhaps more contentiously, any discussion of science policy and policies relating to labour force education, skills and training, despite their potential relevance. These two aspects of policy are important elements of the environment within which innovation takes place, but fall just the wrong side of the divide between general and specific policies that determines whether we address them here or not. Although the state of competition in a market may impact significantly upon the extent of innovation in that market (see, for example, Beneito et al., 2015 for Spain, and our discussion in chapters 7 and 8), we rule out discussion of competition policy per se. Again, it is a matter of the general and the specific. Competition policy in general has many objectives, of which improved innovation is only one and, as such, is not of general interest to us. However, there are some aspects of competition, such as intellectual property rights (IPRs), that are of particular relevance, and we do discuss them. The final major area of policy that we do not address in general terms is trade policy, including tariffs. Thus we do not consider whether free trade stimulates product innovation or not. However, we do consider whether certain trade-related instruments of policy will stimulate product innovation (e.g. foreign direct investment (FDI) incentives, recognition of overseas patents, and IPR).

Having listed what we do not address, it is time to explain what is addressed. We begin by looking at instruments that support the IPRs of innovators. This is an area where there is the most concern for market failure. We look not only at patents but also at design rights, copyrights, and trademarks. We then move to consider other policy instruments, assuming the existence of IPR support instruments. To a large extent, the list of instruments considered has been determined by the constraints to innovation discussed in chapters 7 and 8. The relevance of those different constraints in different countries may help one to understand which policies are of greatest importance in those countries. The list of constraints (excluding those we have decided to not discuss) contains innovation costs, excessive risks, costs of finance, lack of information on, or uncertainty relating to, markets or technology, and regulations. Most of these issues are supply-related, but there are also demand-related issues and policies that may merit discussion as well. The discussions in some cases reflect potential market failures, but in other cases the policies will be justified by the 'more is always better' argument, although resource costs are always an issue when policies are evaluated. We should also state that only in very few cases are policies directed just at product innovation; they may well also impact upon process innovation and other types of innovation. However, as we discussed in chapter 8, firms that are innovative in one dimension tend to be innovative in all dimensions, and thus the distinction may not be that important in practice.

13.2 Intellectual property rights and protection

The World Intellectual Property Organization defines intellectual property as 'creations of the mind, such as inventions; literary and artistic works; designs; and symbols, names and images used in commerce' (WIPO, 2011). The term therefore encompasses the knowledge generated by inventors and innovators that provides the basis for new products, be they physical, intellectual, or creative. IPRs are the administrative instruments provided in law that enable the owners of intellectual property to more effectively control, trade, and charge for the use of the knowledge that they have generated, or at least hold. These powers will better enable them to generate a return from innovation and thus to incentivize their actions.

In chapter 3 we introduced the four main IPR mechanisms: patents, which protect the technical and functional aspects of products and processes; design rights (or design patents), which protect the visual appearance or eye appeal of products; copyright, which protects material such as literature, art, music, sound recordings, films, and broadcasts; and trademarks, which protect signs that can distinguish the goods and services of one trader from those of another. Each has its own particular modus operandi. Given the discussion in chapter 3, we here provide only a brief overview of each.

Patents protect new inventions and cover how things work, what they do, how they do it, what they are made of, and how they are made. They give the owner the right to prevent others from making, using, importing, or selling the invention without permission. In many countries only advances of an industrial nature can be patented, which causes them to be labelled 'utility patents'. In other countries, such as the United States, there is also a form of patents called 'design patents'. These correspond to design rights in many other countries. Once granted a patent, the holder has the right to (not unfairly) determine access to the knowledge embodied in the patent and charge licence fees for use of that knowledge. If granted, in the United Kingdom the patent must be renewed every year after the fifth and may then provide protection for up to twenty years. In the case of infringement the patent holder may take court action seeking termination and damages. Within the EU, firms may apply for either national or European patents, the latter providing community-wide protection, the former only national protection.

Design rights apply to intellectual property in the physical appearance of a product and thus may offer protection to product innovations that would not be covered by patents; they may provide additional protection to the design aspects of a new product over and above that offered by patents. To qualify for protection, designs must be new and individual in character, which means that the overall impression the design gives the informed user must be different from any previous designs. In most countries two main types of design protection are available. In the United Kingdom there are (i) registered design rights, which offer protection throughout the country for up to twenty-five years (although renewal is required every five years), for which application must be made and a fee paid, and which provide, in addition, the right to sell, or licence someone else to use it; and (ii) UK design rights, which is an automatic right that does not need to be applied for, automatically protects a design for ten years after it was first sold or for fifteen years after it was created (whichever is earliest), and prevents others in the United Kingdom from copying; but it covers only the 3D aspects of the item and does not protect the surface decoration of the product or any 2D pattern. At the European level, the equivalents are (i) registered community design right (RCD) and (ii) unregistered community design right.

Copyrights relate to the expression of an idea, not the idea itself or any process by which that idea is embodied in a physical artefact. This contrasts with the patent system, where the idea itself is protected and owned for a period by the patent holder. But it does not protect the names, designs, or functions of the items themselves. Copyright is particularly applicable to new products in the creative industries. Software enjoys only copyright protection and not patent protection. It is not necessary to formally apply or pay for copyright in the United Kingdom.[1] It is an automatic right. The copyright arises as soon as the work is 'fixed', that is, written down, recorded, or stored (in a computer's memory), and in the United Kingdom it is established once the © symbol is attached to the work. The owner of the copyright has the right to license it or sell, or otherwise transfer the copyright to someone else. Copyright in literary, musical, artistic, and dramatic work in the United Kingdom lasts for the creator's lifetime plus seventy years (basically the same as in the EU and the United States). For films it is seventy years after the death of the last of the directors, score composer(s), dialogue or screenplay authors, and for TV and radio programmes it is fifty years from the first broadcast. Sound recording copyright lasts for fifty years. Publisher's right, which covers the typographical layout of published editions such as books or newspapers (presentation on the page), lasts for twenty-five years from creation.

A trademark is a sign that can distinguish a firm's goods and services from those of other traders. A sign includes, for example, words, logos, pictures, or a combination of these. Whereas patents are not available for aesthetic innovations, such innovations may be trademarked. Non-aesthetic innovations may also be trademarked. A registered mark confers the right of use of that mark on the goods and services in the classes for which it is registered, and the legal right to take action against anyone who uses the mark, or a similar mark, on the same goods and services as those that are set out in the registration or on similar ones. To be registerable, the trademark must be distinctive for the goods and services for which application is made and not the same as (or similar to) any earlier marks on the register for the same (or similar) goods or services. A trademark does not have to be registered. An unregistered trademark provides certain rights under common law and the owner can use the TM symbol. However, it is easier to enforce rights if the mark is registered. In the United Kingdom application for registration is made to the Trade Marks

[1] But this is not the case in all countries.

Registry of the Intellectual Property Office. One needs to pay a renewal fee every ten years. European protection via a Community Trade Mark application is made via the European Union Intellectual Property Office (previously the Office for Harmonization in the Internal Market). In addition to national and European trademarks, there is also the WIPO/MADRID International Trademark Registration system. In this system a single application may be made relating to a trademark already registered in a member country for an international registration equivalent to a bundle of national registrations. An international registration is effective for ten years and may be renewed for further periods of ten years on payment of the prescribed fees.

As may therefore be seen, there are several different IPR mechanisms available to owners of intellectual property. The different instruments apply to different types of knowledge and offer different terms of protection, and also different costs. They also offer different periods of protection. A common characteristic, however, is that for all these IPR instruments it is necessary for the owner to personally police his/her rights via the courts, there is no independent policing. Such policing may be costly.

The several IPR instruments made available to innovators are in fact two-sided instruments; patents are especially so. These instruments both protect and reveal. To take advantage of the protection offered, innovators have to declare the knowledge they wish to protect. This declaration has the advantage that it can provide a basis for further advances in knowledge to be made by others and, as such, can prevent unnecessary repetition in the search for knowledge. It is, however, the protection element of IPRs that they are most lauded for. A key issue is thus whether they are able to offer such protection, a possible indicator of this being whether the mechanisms are widely used by knowledge owners; for, although these formal administrative mechanisms exist, there are other means by which owners may protect their intellectual property. One might note, however, that these other mechanisms do not generally carry a reveal advantage and, as a result, society might prefer wider use of the formal IPR mechanisms.

The effectiveness of the different IPR mechanisms is still a matter of research (see, for example, Hall and Harhoff, 2012). In the 2013 UK Innovation Survey (see DBIS, 2016b) businesses were asked how effective they had found a variety of formal and informal protection methods, ranging from patents to secrecy (including non-disclosure agreements), for maintaining or increasing the competitiveness of product or process innovations introduced during the survey period. The proportions reported were low. In the 2015 survey, businesses were asked to rate what proportions of their innovation were protected during the survey period by a variety of given protection methods. The numbers of businesses providing data on protection were rather low. Of the respondents, 10 per cent of all businesses rated complexity as a form of protection for some proportion of their innovation. This was followed by secrecy (including non-disclosure agreements), cited by 7 per cent. Copyright and patents were mentioned by 4 per cent, while design registration was cited by 3 per cent only. At least twice as many large firms cited all the methods as protection for some proportion of their innovation. Hall et al. (2013) note the surprisingly small number of innovative firms that use patents (only 4 per cent), which they attribute to firm size, the preponderance of new-to-firm rather than new-to-market innovations, and the lack of sector patent activity. Pitkethly (2007) reports the results of a survey of IPR awareness and use in 1,700 UK firms in 2006. Firms were asked to indicate the importance to their business of various methods of protecting innovations, both formal and strategic, ranging from unimportant to essential. The non-formal protective mechanisms were seen as at least as effective means of protection as patents and some other IPRs (if not even more effective). Of the formal IPR mechanisms, copyright is considered by most respondents (22 per cent) to be essential, design rights by the fewest (10 per cent), while patents and trademarks rank equally in the middle (for about 13 per cent of firms). Even so, the proportions that consider the formal IPR arrangements as essential are not large. Of the non-formal mechanisms confidentiality agreements were considered essential by 27 per cent of respondents, secrecy by 19 per cent, complexity of design by 7 per cent, and lead time over competitors by 15 per cent. Pitkethly (2007) reports that these data correlate well ($R^2 = 0.75$) with the findings of the 2005 UK Innovation Survey and are broadly similar to findings by Levin et al. (1987) regarding the relative effectiveness of IPRs, lead time, and secrecy. Thus, although not a sine qua non when it comes to innovation, formal intellectual property protection is quite widely used and, in consequence, may well offer more effective protection of knowledge it helps to generate.

The issue that then arises is whether the protection that IPRs offer firms yields a net social gain. The principle behind IPR instruments is that, by strengthening property rights in knowledge, the owners of that knowledge may build up some monopoly power over that knowledge that they otherwise would not have. By exploiting that monopoly power, largely by setting higher prices for use of that knowledge or for the goods in which it is embodied, knowledge owners may generate or increase monopoly profits, and these profits provide an incentive to undertake innovation and thereby stimulate innovation. However, the monopoly power and the possibly higher prices will reduce the level of demand for the knowledge or the goods in which it is embodied. We have shown in chapter 12 that (for prices below the monopoly profit-maximizing price), as the price of an innovation increases, producer surplus increases but consumer surplus declines with the sum of consumer and producer surplus declining. It would thus seem natural that the owners of knowledge support the idea of institutionalized IPRs, but it is not clear that the innovation stimulating benefits will necessarily always exceed the losses to consumers, and thus that such rights are always socially desirable.

For this reason, the length of IPR protection is restricted, in that only for the length of the life of the right is the monopoly protected, and after that period the knowledge may be used freely by all, monopoly profits being no longer artificially enabled. IPR mechanisms thus provide a benefit to the innovator at the cost of lower usage by society as a whole, but only for a specified period. A key issue is the determination of the optimal life of an IPR right. The longer the protection, the greater is the benefit to the innovator, but the smaller is the gain to society. As has been illustrated, the different IPR mechanisms offer different lengths of life, which may also be lengthened in some cases if renewal fees are paid. The determination of optimal lives is a complicated trade-off of costs and benefits and, probably, innovation-specific. It is hard to be convinced that the lives offered by existing mechanisms are optimal, but it is equally difficult to say whether they are too long or too short, and in any case optimal life might well have been changing over time.

For example, there has been much recent discussion of copyrights and their lives. This has happened for three main reasons: (i) with the advent of the internet, there have been increasing amounts of, and increasing concern over, copying and unapproved downloads, especially of music, films, and books; (ii) enforcement of copyright is in the hands of the owner and many owners will be individual artists, authors, or perhaps academics for whom the costs of enforcement are too great; (iii) there has been increasing concern over the issue of orphan works, which are copyrighted works that are inaccessible as the copyright holder cannot be found to approve access and copying. One way to try to address these issues is to change the length of copyright life. Lindsay (2002) argues, for instance, on the basis of the literature, that there is no basis for assuming that the current limits are optimal. It is not clear whether they are too long or too short.

In fact, a number of recent changes introduced into UK copyright law have weakened the power of copyright holders by allowing wider access to copyrighted materials without requiring the permission of the rights holder (IPO, 2014). For example, recent changes enable libraries, archives, museums, and galleries to make copies of all types of creative works in their collections, in order to preserve them for future generations, when it is not reasonably practicable to purchase a replacement. Also, in the EU, the Orphan Works Directive (2012/28/EU) has been introduced, which allows any institution (e.g. public libraries, education establishments, museums, and archives) wishing to use an orphan work to do so after first carrying out a diligent search, in good faith, from appropriate sources.

The consideration of optimal life determination as seeking the right balance between producer and consumer surplus is, however, a rather too simple characterization. There are several reasons for this that, to some degree, have been introduced in chapter 12.

1. Economic welfare can be enhanced by the provision of variety to consumers. It would seem that IPR mechanisms, by limiting emulation, may well limit this proliferation. However, whether it be horizontal or vertical variety that is considered, there is no necessary agreement on whether optimal variety is promoted or inhibited by competition.
2. From an intertemporal viewpoint, there may well be a welfare optimal diffusion path for durable innovations that will be generated by particular intertemporal price paths for new products. However, as shown in chapter 12, whether this path (or the relevant prices) will actually be pursued depends partly upon learning effects, network effects, and the price expectations of buyers, in addition to the monopoly power of suppliers. It is not necessarily the case that a monopoly supplier will produce the welfare optimal diffusion path, and thus whether welfare optimality is promoted or inhibited by the competition or the lack of competition that IPRs promote.
3. Dasgupta and Stiglitz (1980) have argued that, if IPRs have the characteristic that the first to discover wins the right, then this may lead to overinvestment in R&D (the common pool effect), because competitors may compete to be first and to win the prize. In his contribution to Dasgupta and Stoneman (1987), Partha Dasgupta makes the observation that from a normative perspective each inventor or innovator wants to be the winner, whereas society does not care who wins, and thus such races or competitions imply overinvestment from a social point of view.
4. On a similar line to the previous comment, building upon the work of Aghion and Howitt (1992) and Grossman and Helpman (1991), it is also argued in the literature (Jones and Williams, 2000) that, through creative destruction, innovation may lead to a redistribution of rents from past innovators to current. Redistribution, per se, yields no social gain and hence the private payoff to innovation encouraged by IPRs may exceed the social payoff. However, Jones and Williams (2000) consider that, if the appropriability problem (which IPR tries to solve) were eliminated, then R&D would increase by 140 per cent, whereas the creative destruction effect stimulates R&D by only 25 per cent. Thus, quantitatively, it is unlikely that creative destruction can lead to a redistribution of rents from past innovators to current to such an extent that the private payoff to innovation will exceed the social payoff.
5. In addition to the length of life of an IPR, there is the issue of its breadth, that is, the body of knowledge over which exclusivity is claimed, and that the claim may be broad or narrow. The wider the specified right, the harder it is for potential entrants to enter into the rights holder's market. The choice of breadth is that of the applicant, but it is conditioned by the IPR examiner's view. Breadth and optimal life will interact in complex ways.

6. The balance between the costs and the benefits of IPRs is further muddied by foreign trade. The trade-off between consumer and producer surplus is much less clear when the innovations are imported (exported) than when they are produced and consumed in the domestic economy. If a product innovation is imported, IPRs will increase the producer surplus of the overseas producers at the cost of reducing consumer surplus at home. If a product is exported, IPRs overseas will lead to increased producer surplus at home, but the loss of consumer surplus will fall upon overseas consumers. This suggests, first, that domestic welfare can only be improved if exported innovations obtain IPR protection in overseas markets. The greater the length and breadth of this protection, the greater the benefit to the domestic economy. For products that are imported, less protection would seem to be more socially desirable. Of course, no modern developed economy would allow such differentiation of treatment in its IPR regulations between exported and imported products, although for developing countries it is not at all clear that strong IPR rights are beneficial, when most of those rights belong to owners outside the domestic economy.

13.3 Tax incentives to product innovation

In the general run of things, expenditure by companies on the generation and development of new products is deductible against profits, usually in the year in which it is incurred. Such deductibility makes the post-tax cost of product innovation smaller, and thus should assist in overcoming the cost constraint that CIS data indicate is a major barrier to innovation. In addition, the resultant increase in after-tax profits might also alleviate any constraint arising from the availability of funds. Either on the grounds that innovation is good, so more is better, or on the grounds that even with strong IPRs the social returns from innovation are greater than private returns,[2] in almost every country (e.g. in the majority of OECD countries) additional R&D tax incentives have been put in place that provide more favourable treatment for R&D, for tax purposes (either above or below the line), than for other capital investments. Such an instrument is seen as having smaller administrative costs than other instruments and rules out the need for government to pick winners, which may be necessary under project support schemes.

Such tax incentives do, however, have their problems. For example, the size of the incentive would tend to vary as the corporation tax rate varies, being

[2] This is because innovations add to consumer surplus as well as to producer surplus or generate knowledge externalities (or both). Some quantification is provided in Hall et al. (2010) and Bloom et al. (2013).

smaller the lower the rate; tax incentives may be limited if the investor has insufficient profits against which the expenditure can be offset; there is an incentive to classify as R&D other expenditures (e.g. market research) that would otherwise be treated less favourably; and the tax incentive may subsidize activities that would be undertaken in any case. But these problems can often be overcome by appropriate design, for example by defining R&D closely in the rules or by allowing extra tax relief only on increases in R&D. Here, however, we are more concerned with the general than with the specific, each country having its own rules.

The big issue is whether such schemes are effective. There is a wide literature upon this issue; and several surveys of the literature itself are available. That by Kohler et al. (2012) takes an international perspective and is accessible. Dechezleprêtre et al. (2016), in addition to providing a review of the current state of this literature, is a significant new contribution that not only analyses the impact of a change in the rules of the UK scheme—which gave more favourable treatment to small and medium enterprises (SMEs) than was previously given—to better estimate the impact the impact of R&D tax credits, but explicitly considers that the effectiveness of such schemes should be judged in terms of impact on innovation and not in terms of R&D itself, for R&D is the input to generating innovation and not the output. The authors measure innovative activity by exploring the impact of tax credits on patenting activity.

Dechezleprêtre et al. (2016) find large effects of the UK tax credit scheme on R&D and patents. They estimate an elasticity of R&D (with respect to its tax-adjusted user cost) of about 2.6—a value higher by between one and two than those typically estimated in the recent literature, a result put down to the small firm bias in the sample and the policy change analysed. To illustrate the impact of R&D tax credits in the United Kingdom they estimate that between 2006 and 2011 the UK R&D Tax Relief Scheme induced £1.7 of private R&D for every £1 of taxpayer money and that aggregate UK business R&D would have been about 10 per cent lower in the absence of the policy. On the crucial issue of whether innovation—the main objective—is also increased, they also find that, as a result of the policy change, patenting rose by about 60 per cent, with no evidence that the innovations were of lower value.

One possible side effect of R&D tax credits is that, by raising the R&D activities of all firms, one reduces the potential gross payoff from an innovation as a result of increased competition, which may in part (or, in the extreme, in whole) reduce the net return to a previously profitable innovation and deter from some activities that would have otherwise taken place. Dechezleprêtre et al. (2016) find, however, that the policy-induced R&D has sizable positive impacts on the innovative activity of not only R&D performing firms but also other firms in similar technology areas.

Such results do tend to provide strong support for the use of tax incentives (properly designed) as a means of increasing R&D and patenting, and thus product innovation in an economy. It should be noted, however, that some of the increased R&D may reflect a transfer of activity from other economies with less generous tax incentives, and not an increase in the worldwide search for knowledge. Although it might appear that it makes little difference where R&D is undertaken, there is some evidence that the country where R&D is performed obtains a disproportionate share of its productivity benefits, at least initially (Jaffe et al., 1993). There may therefore be local gains from an inward transfer of R&D activity, even if world-innovative activity does not increase. One might also argue that externalities or spillovers from scale can be seen as desirable. However, if policies are mainly going to impact via their relocation effects, then clearly international trading rivals are not going to stand idly by; there is thus the potential of R&D tax credits being subject to a race to the bottom.

It is of interest that the possibility of relocating innovative activity has been explicitly pursued in several countries via an alternative tax subsidy scheme—the so-called patent box. This is a scheme that exists (or has existed) in various forms in different countries, for example Belgium, France, and Spain, since being first introduced in Ireland in 2000, and was introduced in the United Kingdom in 2013. In the United Kingdom the patent box allows a 10 per cent tax rate, a rate considerably below the current rate of corporation tax, on profits derived from any products that incorporate qualifying patents (encompassing patents granted by an approved patent-granting body such as the UK IPO, the European EPO, and various designated European territories, but not the United States, France, and Spain). The patent box excludes products that have only copyright or trademark protection. The initiative is specifically aimed at stimulating the exploitation of patented knowledge (rather than its generation) in the United Kingdom, wherever it was generated. The steady state cost after the initial phasing-in period of the patent box was forecast to be approximately £1.1 billion in terms of corporation tax revenues foregone. This was seen as a measure that would incentivize commercialization of intellectual property in the United Kingdom. It was also seen as a logical counterpoint to changes in the tax regime that covered controlled foreign companies, which had made it harder for companies to take their profits offshore, by rewarding those companies who bring their intellectual property profits into the United Kingdom.

However, not surprisingly (given the largely zero-sum game involved), the scheme led to considerable international disharmony and especially to complaints from Germany, and the existing UK regime was closed to new entrants for both products and patents in June 2016 (the existing scheme will be abolished in full by June 2021). Fundamentally, patent-box relief will now be restricted to profits generated from intellectual property initially developed in the United Kingdom, although intellectual property within the existing regime will be able to retain the benefits of the scheme until June 2021, to allow time for transition to a new agreed (nexus) regime. This might increase the pressure to incentivize R&D transfer via R&D tax incentives.

R&D tax incentives, as the name implies, of course relate only to R&D. For purposes of credit, R&D is defined largely in line with the OECD (2002) definitions, as consisting of activities that seek to achieve an advance in science or technology. However, such expenditures are not of great significance in a number of industries such as the service sector or the creative industries, and are also different from expenditures on design. Much expenditure on product innovation may thus lie outside these schemes. Partly offsetting this apparent bias, a new set of tax reliefs has recently been granted to certain creative industries in the United Kingdom. There is now a group of six such reliefs: film tax relief was introduced in April 2007; animation tax relief, high-end television tax relief, and children's television tax relief were introduced in April 2013; video games tax relief and theatre tax relief were introduced in 2014. These operate in a somewhat similar way to R&D tax credits. However, they appear to have more of a grounding in trying to create a cultured society, in addition to being driven by issues of international competitiveness; for, to qualify for these creative industry tax reliefs, all films, television programmes, animations, or video games must pass a 'cultural test' or formally qualify through an internationally agreed co-production treaty—certifying that the production is British. The United Kingdom is not the only country to offer such incentives. For film production, for example, the Italian government since 2009 has given local producers a 20 per cent tax break, outside investors a 40 per cent tax shelter, and foreign productions a 25 per cent tax credit. In the United States there are various state-level incentives to encourage film-makers; these encompass for example tax credits, cash rebates, and grants. In many other countries similar schemes are also available (see, for example, www.filmcommissioners.com/Incentives-WesternEurope.html).

The one classic argument against R&D tax-support programmes is that, almost by definition, they lead to deterioration in the public sector borrowing requirement (PSBR), at least in the short term. At a time when control of the PSBR is often considered a major characteristic of correct fiscal policy, this may be politically undesirable. An alternative is therefore what is called a 'levy grant system'. In such a system, all firms in an industry pay a levy that is then redirected to those companies that are undertaking R&D or being innovative. This clearly has an effect on the costs of innovation, and also on the availability of finance. It is a type of scheme that has been used in the past to finance training schemes in the United Kingdom and is currently being reintroduced to finance apprenticeship schemes. The theoretical foundation supporting such schemes can be found in Stoneman (1991).

In addition to (potentially enhanced) tax relief on R&D, businesses are generally allowed to charge the costs of various specified capital expenditures against profits prior to paying tax. Governments attempt to incentivize such investments through manipulation of the rules on the times at which the expenditures can be charged and the rates of relief offered. As most capital will last for a number of years, bringing forward the allowed date of offset will encourage investment, as will higher rates of offset. Such policies are, again, widespread and will impact on product innovation in two ways. The policies will reduce the cost of tooling up to produce product innovations and will also help in alleviating any financial constraints. They will thus stimulate the supply side. In addition, if one is considering the demand for a product innovation that is a producer durable, then a favourable tax treatment will encourage demand for that innovation. That encouragement could also speed diffusion via demonstration effects. However, diffusion subsidies might be offset, inter alia, by price increases or expectational effects.

13.4 **Government grant-based policies**

The main policy alternative to tax reliefs pursued by governments as a means to stimulate innovative activity is the use of project subsidy or finance schemes to which firms may apply for funding in order to undertake specific innovation projects (for some theoretical support for such schemes see, for example, Herguera and Lutz, 2010). These schemes are widespread (see Takalo et al., 2013) and are exemplified by the South African Support Programme for Industrial Innovation (SPII), a programme relaunched in 2015, which was designed to promote technology development in South Africa's industry by providing grants to firms for the development of innovative products or processes or both. Dimos (2016) provides a survey of the effectiveness of R&D subsidies covering fifty-two microlevel studies published since 2000. It might, however, be argued that, to some extent, tax incentives to R&D have become more popular (the United Kingdom terminated its scheme of product and process development grants some years back).

Such schemes should lower the marginal cost of innovative activity and thus relax the cost constraint on R&D, but the availability of public finance may also help to overcome any funding availability constraint. Funding via such schemes may also have a reputation effect on those so supported and be desirable to firms for just that reason. Of course, there is a cost to society from such schemes, in terms of both the funds deployed and the administration involved (which may be a reason for their reduced popularity), and so there is considerable interest in whether they are effective. It is often argued that government cannot pick winners. The effectiveness of these schemes has often been considered in terms of whether they provide additionality, which is largely measured by whether the public R&D leads to an additional amount of private R&D.

David et al. (2000) survey the earlier literature relating to the effectiveness of such schemes but, for very valid reasons, do not arrive at any definitive empirical conclusions regarding the sign and magnitude of the relationship between public and private R&D. More recently, Takalo et al. (2013) have addressed the issue in the context of a Finnish R&D subsidy program. They find that the subsidy scheme on average produces significant welfare gains, about 60 per cent of the gain being internalized by firms and the balance going to society more widely. They calculate that the expected programme benefits exceed the opportunity cost of the public funds spent.

Although the discussion of technology-based innovation and the use of subsidies is most widespread, subsidies may also be used to encourage innovation in the creative industries, and especially the performing arts. In England, the Arts Council is the primary body that invests in the arts, providing subsidies to theatres, opera companies, and orchestras. The Council states that 'arts and creativity will continue to play a significant part in injecting innovation and enterprise into the economy' (www.artscouncil.org.uk) and will contribute to developing and sustaining the creative economy by (i) funding for-risk investment in new work and new talent that stimulates connections between the subsidized and commercial creative industries; (ii) supporting arts education activities in order to foster creative thinking at all life stages; (iii) funding R&D that links arts with other aspects of the economy, such as industry and science; and (iv) investing in new business models, leadership development, and partnerships designed to develop creative clusters and build regional prosperity and sustainable communities. Note that this very much appears to be framed in terms of spillovers and potential market failures.

13.5 **Public contracts and government acquisition**

Government not only stimulates innovation through its tax and grant support schemes but also is a major customer for new and existing products and may in various ways use this purchasing power to influence the innovativeness of its supply base (for an overview, see Uyarra et al., 2014). This is nowhere more obvious than in the defence sector, where the government is probably the only domestic customer and its needs and demands will significantly affect new product development. Although the government may consider that defence spending has the primary objective of defending the nation, its procurement decision may well have wider impacts. These may be considered at a number of levels.

The choices made between development and production at home and the importation of defence equipment could have a vital role to play in whether new (defence) products are made in the domestic economy or not. In addition, the specification for products to be developed and produced at home could significantly impact upon later defence sales and exports.

Clearly the government demand will (partly or wholly) lead to product innovation that would not have otherwise occurred. Depending on the nature of the contracts employed in defence procurement, government activity in this area may or may not lead to an overall increase in domestic innovation beyond what it would have been if the resources had been used elsewhere. The counterfactual is difficult to evaluate. It is also the case that the innovation spinoff from domestic development and production can lead to knowledge advances in the domestic workforce and in the domestic economy, yielding wider benefits. However, development and production at home can make considerable demands upon the existing skilled workforce, reducing its availability for other activities (or increasing its costs) and reducing the wider spinoff benefits from these activities. Latest estimates from Moretti et al. (2016), however, show strong evidence of crowding in rather than crowding out, as increases in government-funded defence R&D result in significant increases in private-sector R&D. Specifically, a 10 per cent increase in government-financed R&D generates about 3 per cent more privately funded R&D. This increase in privately funded R&D is shown also to lead to significant increases in R&D employment and to sizeable productivity gains.

Much product innovation in free market economies requires firms to develop products at their own expense and then to place that product on the market. If successful, the product will return a revenue that will in time more than cover the costs of development. This involves investment, risk, and uncertainty, with no guarantee that customers will exist or that they will be willing to pay an appropriate price for the innovative products. Although there are some private sectors where innovation is undertaken under contract to government bodies (e.g. large construction and infrastructure projects), large parts of defence procurement are undertaken using development (and production) contracts, which after an initial possible competition for the contract offer an environment where product innovation can occur with at least some of the future market guaranteed, and thus with much less uncertainty and market risk.

The risk carried by the innovating suppliers will to some degree be affected by the terms of the contracts issued by government. It used to be the case that much defence procurement was undertaken on a cost-plus basis. Government paid the cost of the project plus a profit margin. It was clear, however, that such contracts did not encourage (cost) efficiency. It also seemed to be the case that such contracts did not imbue contractors with attitudes to development that could have been beneficial in the more open non-government market place; nor did they encourage spinoff activity. Now most contracts are specified in fixed-price terms, encouraging greater efficiency. This tends, however, to make the costs of changes of specification a bone of contention once a contract has begun. Such defence contracts will often have stage payments that provide the developer with finance during the development process. This can significantly alleviate any finance constraint that the developer may otherwise face.

It is not clear whether (all or some) such procurement contracts specify who owns the IPRs on the innovations made. If they are owned by the innovators, then those innovators may be able to exploit that intellectual property outside the defence sector or by supplying products to other customers (although such supply may be regulated in various ways, for political reasons). If the innovators have some rights over the intellectual property, then government procurement might be seen as a route by which the development of product innovations for other markets can be encouraged. Such issues may be of particular relevance if the development contract is placed with an overseas supplier, for, although the government may meet its defence objectives, there will be no necessary improvement in the domestic supply base and, moreover, the intellectual property is more likely to be located overseas. One must also note that, despite the costs of defence procurement, it is not at all clear that much of the activity in this area (not just in the United Kingdom, but in other countries as well) has successfully delivered the desired innovations on time (see, for example, Defence Committee, 2004).

Of course, not all procurement is in the defence sector. Contracts in the non-defence sector would include large intellectual property projects that are designed to supply large innovative computer systems to government and involve the development of complex software. The suppliers are often but not always overseas-based. There is, however, a fear that such schemes are no more successful than defence projects of the same kind; for example, in 2004 the National Audit Office in the United Kingdom reported: 'The history of such procurements has not been good, with repeated incidences of overspends, delays, performance shortfalls and abandonment at major cost' (National Audit Office, 2004, p. 1). Such experiences continue. An alternative example has been the recent government negotiations to acquire for the United Kingdom its first nuclear power plant for a generation. In this case the government has not provided funding as such but has provided guarantees over the future price that the owner of the plant will receive for the electricity supplied. These two cases (IT acquisition and nuclear energy) are also of interest because the suppliers are generally overseas companies (French and Chinese, in the case of nuclear energy) and thus, although the innovation is aimed to be for the benefit of the British consumer, the support will be for product development overseas. It is a rare case where the support programme is not aimed at improving home capacity to innovate and produce but rather to increase the flow of benefits to consumers in the United Kingdom from product innovations whose development is based overseas.

Other examples of non-defence procurement can also be found. As the main provider of health services, the government in the United Kingdom has significant bargaining power with pharmaceutical companies. Via the National Institute of Health and Care Excellence (NICE), it uses this power to do more than determine which new drugs and treatments will be funded by the National Health Service (NHS): it can often have considerable impact upon the price of new medicines to the NHS. This is an example of demand-side intervention, as opposed to the supply-side interventions mainly discussed, and such intervention can affect considerably the diffusion of a product innovation. Another example of how government can use its buying power to affect domestic product innovation is the Pharmaceutical Price Regulation Scheme (PPRS), which has existed in various forms since 1957. In its current form it is a voluntary agreement between the Department of Health (DH) and the Association of the British Pharmaceutical Industry (ABPI) regarding the supply and especially the cost to the NHS of branded medicines across the four nations of the United Kingdom; and it regulates the profit that companies can achieve. At the launch of the current scheme in 2014 the secretary of state expressed the view that the PPRS offered an opportunity to agree and carry through a solution for accelerating uptake of clinically and cost-effective medicines. This is a clear example of government purchasing power being used to stimulate the take-up of product innovations. In a similar way, government purchasing power may also be used to encourage the spread of product innovations via demonstration effects. A technology may be used by government at an early stage in its life partly in order to inform the market as to its potential and thus encourage wider diffusion. Alternatively, government could fund demonstration projects in private firms with a similar effect.

13.6 **Addressing financial constraints**

Although a number of the polices already discussed will help to relieve financial constraints on innovation, that result is often a side effect to a main policy objective of reducing the costs of innovation. However, because external financing is a way of shifting risk, relieving constraints on firms that are financially constrained in their innovative activity may also reduce the riskiness of innovation. There are, additionally, specific government policies that are aimed at reducing such risk. One method is for government to subsidize financiers who provide funding for start-up or technology companies where risks are high. Two such schemes in the United Kingdom are the Enterprise Investment Scheme (EIS) and the Seed Enterprise Investment Scheme (SEIS), which give income and tax relief on capital gains to individuals who invest in small early-stage businesses. Similarly, private investors in venture capital schemes in the United Kingdom can obtain tax relief on their investments and relief from tax on capital gains.

More significant, however, is a scheme known as launch aid. First introduced in the United Kingdom in 1947, this is a loan scheme through which government lends funds to companies to finance product development; and that loan, plus interest at a predetermined rate r, is paid back from sales of the product developed, usually as a charge on sales above a number n up to a limit m. The theoretical basis of the scheme is discussed in Kaivanto and Stoneman (2004, 2007), and the details of its operation in Kaivanto (2006). When first introduced in the United Kingdom, this policy was designed as an aid to the civil aerospace industry, to cover the development of new civil aircraft and aero engines. The rationale given for its existence was that such projects required large investment and were very risky, making it difficult to raise finance from the private market. The scheme has been extended to many other EU countries and has been extensively used to support the development of new products by Airbus.

That the scheme could relax a financial constraint is obvious. But there has also been a long-running dispute between the US aerospace industry (especially Boeing) and EU governments over whether the scheme also provides illegal subsidies to EU industry, and especially to Airbus. Whether a subsidy element is also involved will depend on the interest rate charged for the loans and on whether n and m are set such that repayments will be made on sales below an acceptable minimum and up to a sales target that is sufficient to recover the loan and interest under a realistic sales scenario. Boeing argues that launch aid provides significant advantages to Airbus, in that the interest rates on the loans are significantly less than what commercial lenders would charge and, in the event that a product does not hit a predetermined sales target, the remaining loans on the product are forgiven. This provides an artificially low cost of capital, lower programme risk, and the ability for launch aid to price its products lower than the competition. It also enables it to introduce new products faster than it would be able to do otherwise. In response to a US complaint, in June 2010 the World Trade Organization, WTO, ruled that Airbus had received $18 billion of illegal subsidies, including $15 billion of launch aid (http://www.boeing.com/company/key-orgs/government-operations/wto.page). It would thus seem that launch aid did meet its target of stimulating the EU aerospace industry (much to the chagrin of competitors), but the terms of the aid provided were determined to be too generous.

Launch aid-type schemes have also been used outside the aerospace industry, but only to a limited extent, for example in the Netherlands and Finland, to support SMEs. It is surprising that they have not been more widely used.

13.7 **Access to knowledge**

Another reported constraint on innovation in firms is knowledge shortages on the part of potential innovators. Policies designed to facilitate the spreading of knowledge may alleviate this problem. Examples in the United Kingdom include schemes led by Innovate UK, an independent body with its origins within the Department of Trade and Industry. Early attempts at technology transfer organizations were known as Faraday Partnerships and based on a previously successful German model (the Fraunhofer Society in Germany, which is Europe's largest and most successful organization for applied research and technology transfer). Currently, central to the programme are the so-called Catapult centres, which are independent not-for-profit physical centres that connect businesses with the United Kingdom's research and academic communities in specific areas where businesses, scientists, and engineers work side by side on late-stage R&D, transforming high-potential ideas into new products and services. Each Catapult centre is expected to raise funds equally from three sources: business-funded R&D contracts; collaborative applied R&D projects from the United Kingdom and Europe, funded jointly by public and private sectors and won competitively; and core UK public funding. Rombach (2014) provides a discussion of the effectiveness of these Fraunhofer-type technology transfer mechanisms.

13.8 **Infrastructure**

Although not always requiring government expenditures, in most economies investments in infrastructure require at least government approval and in many cases will involve public decisions on form, design, and funding. Such decisions can have major impacts either on the development or on the spread of product innovation. Two examples illustrate the case. With the advent of mobile telephony, major decisions had to be made about the (re)allocation of bandwidth spectra. Without this allocation, mobile telephony could not proceed. As time goes on, further allocation of bandwidth becomes necessary, and thus infrastructure decisions of this nature are still of crucial importance. For example, different bands of spectrum are able to transmit more data than others, and some bands transmit more clearly than others. In the United States the Federal Communications Commission (FCC) shifted the location of television on the spectrum, in order to open up more space for mobile applications. The allocations of frequency to operators may occur in different ways. In the United Kingdom the 3G spectrum was allocated via auctions, through which the mobile phone operators paid £22.5 billion (2.5 per cent of GNP) to the UK treasury (Binmore and Klemperer, 2002).

A second example concerns the provision of charging stations for electric cars. Governments have become involved in this roll-out. For example, the Dutch government initiated a plan to establish over 200 fast-charging stations across the country by 2015, aiming to provide at least one station every 50 kilometres (31 miles) for the Netherlands' 16 million residents. According to a press release issued by the UK government in February 2013 (https://www.gov.uk/government/news/new-measures-announced-to-support-the-uptake-of-plug-in-vehicles), the UK government agreed to provide 75 per cent of the cost of people installing charge points where they live (with a total cost to government of £13.5 million); local authorities installing rapid charge points to facilitate longer journeys, or providing on-street charging on request from residents who have or have ordered plug-in vehicles (£11 million); train operators installing new charge points at railway stations (£9 million); and charge points on the government and wider public estate to be installed by April 2015 (support of up to £3 million). The justification for such spending is that the greening of the economy is seen as socially desirable and unlikely to occur, without such assistance, at the rate the government wishes to see.

13.9 **Regulation**

As in infrastructure involvement, government may have major impacts upon product innovation via its regulatory policy. Such policies may encourage or hinder specific innovations. A particular example is the control of exhaust emissions. The first legislated emission controls were introduced in California for the 1966 model year, for cars sold in that state, followed by the United States as a whole in model year 1968. Over time, standards have been progressively tightened. EU Regulation No 443/2009 set an average CO_2 emissions target for new passenger cars of 130 g/km in the EU, a target gradually phased in between 2012 and 2015, and a target of 95 g/km will apply from 2021. For light commercial vehicles, an emissions target of 175 g/km will apply from 2017, and 147 g/km from 2020.

Such regulations will clearly provide direction to future product innovations and are primarily designed to alleviate the impact of motor vehicles upon the environment. It is sometimes argued that such regulations may also (i) stimulate product innovation where it would not otherwise have happened and (ii) encourage the development of innovations that could provide a foundation for more successful future international trading.

As an alternative, but rather less important, example of how regulation can limit (the spread of) product innovations, the Segway is rarely seen in public in the United Kingdom. It (and self-balancing scooters) are not allowed on UK roads and pavements, under section 72 of the Highway Act 1835 in England

and Wales and under the Roads (Scotland) Act of 1984 in Scotland. This is an example in which government regulation is undertaken for perceived health and safety purposes. Other examples are electrical safety and chemical transportation regulations. Such regulations can make potential adopters of new products feel that the products are of higher quality, or at least not life-threatening, and thereby encourage their adoption. The regulations may, however, make the products more expensive and thus limit their diffusion.

There are at least two other ways in which regulations regarding standards can impact on the development and diffusion of product innovations. First, new products often come initially in several different, incompatible forms, for example different video recorder formats such as VHS and Betamax. Although the market may itself settle upon a dominant format (not necessarily the best), the process may generate purchaser uncertainty and slow diffusion. Government selection of a dominant format might therefore be suggested. To have any impact, the selection process will have to come early in the life of the product, when knowledge bases are weakest and thus there is little confidence that the government would make a choice of the best technology.

The third role that standards regulations may play concerns interconnected technologies. When new products have several interconnecting parts, potentially from different suppliers, there may be a social gain to the fact that the parts from the different suppliers are capable of interconnection and interchangeable. Government may intervene to select appropriate interconnecting standards and to impose those standards upon suppliers. For further discussion of standards, see David and Greenstein (1990) and Blind (2013).

13.10 Foreign direct investment

Global companies need not necessarily produce new products in the economies in which they have been generated (see, for example, the discussion of Apple in chapter 6). A typical example is that large global car companies may locate the production of their new models in quite different economies from those in which the models in question are designed and developed. Governments may wish to encourage such production in their domestic economy (and perhaps deter investment in the other direction) insofar as FDI in the domestic economy is a means of generating employment, output, and trade advantages. FDI may also offer spillover benefits to other parts of the economy. Of course, there are other attitudes to FDI, as it is often argued that FDI is a means by which transnational corporations exploit the resources of developing countries to the advantage of the corporations rather than of the host countries. A balance between the two is often sought; for example, China has had considerable success in retaining local ownership of its economy while encouraging technology transfer via FDI. The United Nations Conference on Trade and Development (UNCTAD) (2017) reports global FDI flows in 2016 of $US1.5 trillion. (http://unctad.org/en/PublicationsLibrary/webdiaeia2017d1_en.pdf).

Emphasising the benefits to domestic product innovation that FDI brings without denying the potential downsides (for empirical evidence for the link between FDI and product innovation, see Lin and Lin, 2010), here we concentrate upon efforts designed to encourage FDI. Extensive literatures indicate that the host-country characteristics that attract FDI include market size, access to overseas markets, tax regimes (especially the potential to export profits), skill availability, university quality, capital markets, infrastructure, and regulations. The relative importance of these factors will differ by country. Many of the government policies that we addressed elsewhere in this chapter, such as tax policy, regulations, and subsidies, may, as a spinoff, also encourage FDI without being so directed. Policies aimed directly at FDI may also exist, although the instruments may not be much different from those employed with other intentions.

In the United Kingdom a body called UK Trade and Investment has the brief to encourage FDI in the United Kingdom and works with overseas investors looking to this country as an investment destination. Although the United Kingdom is particularly successful in attracting FDI, no special or additional incentives are on offer. In Ireland, too, FDI has been extensive: 174,000 people were reported in 2015 as employed in foreign-owned enterprises in Ireland, representing almost one in ten workers in the economy. These enterprises also generated exports of €124.5 billion from Ireland in 2013 and €1.4 billion spent on R&D (http://www.idaireland.com/docs/publications/IDA_STRATEGY_FINAL.pdf).

The Irish Development Agency (IDA) states that, for companies that make the decision to invest, IDA Ireland provides financial support in the form of employment, capital, R&D, and environmental or training grants within EU state aid rules. The tax regime (on which see http://taxsummaries.pwc.com/uk/taxsummaries/wwts.nsf/ID/Ireland-Corporate-Tax-credits-and-incentives) encompasses (i) a 12.5 per cent corporation tax rate on active business income, a 25 per cent credit on qualifying R&D expenditures, and thus a total effective tax deduction of 37.5 per cent; (ii) the ability to exploit intellectual property at favourable tax rates; (iii) accelerated tax depreciation allowances for approved energy-efficient equipment; (iv) the ability to carry out investment management activities for non-Irish investment funds without creating a taxable presence in Ireland for such funds; and (v) an effective legal, regulatory, and tax framework to allow for the efficient redomiciliation of investment funds from traditional offshore centres to Ireland. As an example, Ireland has a favourable tax regime for entities known as 'Section 110' companies. A Section 110 company is an Irish resident special-purpose company that holds or manages (or both) 'qualifying assets' and satisfies a number of conditions. A Section 110 company can provide an onshore investment platform, which should be eligible to access Ireland's double taxation treaties (DTTs) network, where the Irish company is the beneficial owner of the income flow. The Section 110 regime has been in existence for almost twenty-five years and, with appropriate structuring, can provide for an effective corporation tax rate of close to 0 per cent. The regime is widely used by international banks, asset managers, hedge funds, private equity firms, and investment funds in the context of securitizations, investment platforms, collateralized debt obligations (CDOs), collateralized loan obligations (CLOs), acquisition of distressed loan portfolios, big ticket leasing, and capital markets bond issuances. The extension of the Section 110 regime to include plant and machinery has particularly increased the attractiveness of Ireland as the preferred destination for aircraft financing and leasing activities.

13.11 Subsidies to the demand for product innovations

Particular product innovations have, over the years, enjoyed public subsidies to their use or ownership. The products may be producer products or consumer products. For example, in the United Kingdom computer ownership by business and in schools has at times been subsidized. The purpose was to encourage computer ownership and use, seen as a significant contribution to the United Kingdom's competitiveness, while potentially encouraging innovation in the domestic computer industry. Currently green objectives underlie the subsidization of other technologies. For example, the UK government provides a subsidy of up to £5,000 for the purchase of eligible electric cars and up to £8,000 for eligible electric vans. This subsidy has been widely claimed.

Another such subsidy has been for the installation of solar panels on UK domestic properties. This scheme has two parts. First, there is a payment from the government to householders for all the electricity they generate, whether they use it or not. It's guaranteed for twenty years and is index-linked, so it rises with inflation This rate is currently (October 2016) 4.18p/kWh, a rate much lower than it had been prior to February 2016, when it was 12.03p/kWh. Secondly, there is an export tariff that is a payment for energy not used by the household itself and sent back to the grid of 4.91p/kWh. (It is estimated that a typical house can earn £85 per annum through the export tariff.) Although the subsidy has considerably reduced over time, it is estimated that the price of a typical solar panel system, including installation, is now around £5,000–£8,000, whereas earlier such a system would have cost £10,000–£12,000. Even so, such a system over twenty-five years could generate a subsidy payment of over £25,000.

The solar panel subsidy is in fact an example of why the subsidization of particular product innovations may not be optimal. The fact that the cost of the panels has fallen over time implies that the subsidy scheme in fact encouraged the introduction or diffusion of this new technology when it was particularly inefficient and expensive, and it might have been better to wait for a cheaper better technology to be developed. The response to this criticism is that, by subsidizing early users, the government provided exemplars that may be followed by others. Also, by stimulating demand, it encouraged the further technological developments that allowed the cost to fall.

One might note that most solar panels in the United Kingdom have been (and are) imported. The subsidy scheme is thus not necessarily encouraging the production of product innovations in the domestic economy. It is very much a demand-side-driven policy. Such demand-oriented policies are not restricted to functional innovations. For example, in most countries there are subsidies to the arts (e.g. opera, the theatre, or classical music) that enable new works to be more widely enjoyed; the subsidies may enable the price of theatre or opera seats to be reduced. Such policies seem reasonably widespread; however, there are few data that illustrate how, for example, the subsidy to an opera company is actually used to reduce seat prices as opposed to covering other expenses, such as raised fees for performers or the costs of better staging. It is reported that in Cologne the City Treasury subsidizes each opera ticket by €165, the Berlin Staatsoper enjoys a subsidy of €248 per ticket, the Hamburg State Opera tickets are subsidized by €114, and the Bavarian State Opera has a subsidy of €90 (http://intermezzo.typepad.com/intermezzo/2012/01/the-real-cost-of-german-opera-tickets.html, accessed 13/6/16). This, however, is only a measure of the subsidy divided by the number of seats, it is not a measure of the extent to which each seat price is lower than it would have been in the absence of the subsidy.

13.12 **Prizes**

The final policy instrument that we shall discuss is one of the oldest. Prizes may be used by government to encourage either product innovation in general or a direction of innovation in particular. The classic historical example is the search for a means of measuring longitude at sea. The Longitude Act 1714 offered a series of rewards, dependent on accuracy, for anyone who could find a practical way of determining longitude at sea. The administrators of the prize, the Board of Longitude, met in February 1765 and decided that the method (accurate timekeeping at sea) represented by John Harrison's watch had worked within the most stringent limits of the 1714 Act, its error being just 39.2 seconds or 9.8 miles (15.8 km) at the latitude of Barbados. Their recommendation was that parliament award Harrison (1) £10,000 when he demonstrated the principles of the watch, and (2) the remaining £10,000 (minus any payments already made) once it was shown that other makers could produce similar timekeepers. The Harrisons felt that the full reward was already due under the terms of the 1714 Act and that the commissioners had unfairly changed the rules (see http://www.rmg.co.uk/discover/explore/longitude-found-John-Harrison#tmZf709kLTozzH9X.99).

In more recent times, prizes have been used most noticeably in architectural competitions to encourage innovative ways of meeting certain design criteria (see Ronn et al., 2013). Examples include the Sydney Opera House, the Guggenheim Museum Bilbao, and the Millennium Bridge London. The downside of competitions of this kind is that the losses that non-winners may bear can deter entrants and perhaps require increased charges elsewhere in the system. As in the case of patents, it may be that competitions of the 'winner takes all' type give rise to the common pool problem and cause overinvestment in designs. On the other hand, the process has generated some dramatic innovations in the past.

13.13 Conclusions

In this chapter we have addressed why governments intervene in the product innovation process and offered three reasons: more is better; international comparisons; and market failure. We do not necessarily agree that these are always rational grounds. After a further discussion of institutional IPRs and the optimality of their respective lives, we moved to consider various other policies that have been employed. This discussion takes on board the finding in chapter 8, that firms that are innovative in one dimension tend to be innovative in others, and that, in consequence, what often matters for policy is whether it promotes innovative firms rather than single innovative activities. Bearing this in mind, the discussion is undertaken in the light of constraints to innovation discussed in chapter 8. A number of policies (e.g. tax incentives and stimulation of FDI) aim to stimulate the generation and production of product innovations in the domestic economy. Such policies, to some degree at least, may attempt to transfer these activities from competing economies; hence they are subject to international rules and may generate international disputes. Other policies may operate on the demand side, stimulating the use of product innovation in the domestic economy regardless of whence they come. In these discussions we have tried to resist the temptation to think of R&D as the only input to innovation (design and creative activities also matter) and to also consider non-technological activities (e.g. innovation that is aesthetic rather than functional).

REFERENCES

W. J. Abernathy and J. M. Utterback (1978), 'Patterns of industrial innovation', *Technology Review* 80 (7), 40–7.

D. Acemoglu (2009), *Introduction to Modern Economic Growth.* Princeton, NJ: Princeton University Press.

R. Adams, J. Bessant, and R. Phelps (2006), 'Innovation management measurement: A review', *International Journal of Management Reviews* 8 (1), 21–47.

Ó. Afonso (2011), 'R&D direction and north-south diffusion, human capital, growth, and wages: A review article', *Economics Research International* (Article ID 401928). http://dx.doi.org/10.1155/2011/401928

P. Aghion, M. Dewatripont, and J. C. Stein (2008), 'Academic freedom, private-sector focus, and the process of innovation', *RAND Journal of Economics* 39 (3), 617–35.

P. Aghion and P. Howitt (1992), 'A model of growth through creative destruction', *Econometrica* 60 (2), 323–51.

H. Ahnert and G. Kenny (2004), *Quality Adjustment of European Price Statistics and the Role for Hedonics* (Occasional Paper Series, 15). Frankfurt: European Central Bank. https://www.ecb.europa.eu/pub/pdf/scpops/ecbocp15.pdf

S. Alänge and A. Steiber (2011), 'Diffusion of organisational innovations: An empirical test of an analytical framework', *Technology Analysis and Strategic Management* 23 (8), 881–97.

E. W. Anderson and M. W. Sullivan (1993), 'The antecedents and consequences of customer satisfaction', *Marketing Science* 12, 125–43.

D. Añón Higón and P. Stoneman (2011), 'Trade in final goods and the impact of innovation', *Economics Letters* 110 (3), 197–9.

C. Antonelli (1985), 'The diffusion of an organizational innovation', *International Journal of Industrial Organization* 3, 109–18.

A. Antoniades (2015), 'Heterogeneous firms, quality, and trade', *Journal of International Economics* 95 (2), 263–73.

K. Arrow (1962), 'Economic welfare and the allocation of resources for invention', in R. R. Nelson (ed.), *The Rate and Direction of Inventive Activity* (NBER). Princeton, NJ: Princeton University Press, 609–26.

K. Atuahene-Gima (1996), 'Market orientation and innovation', *Journal of Business Research* 35 (2), 93–103.

D. B. Audretsch, D. F. Kuratko, and A. N. Link (2016), 'Dynamic entrepreneurship and technology-based innovation', *Journal of Evolutionary Economics* 26 (3), 603–20.

A. D. Bain (1962), 'The growth of television ownership in the United Kingdom', *International Economic Review* 3, 145–57.

P. Baines (2014), *Marketing* (3rd rev. edn). Oxford: Oxford University Press.

W. E. Baker and J. M. Sinkula (2009), 'The complementary effects of market orientation and entrepreneurial orientation on profitability in small businesses', *Journal of Small Business Management* 47 (4), 443–64.

G. Ballot, F. Fakhfakh, F. Galia, and A. Salter (2015), 'The fateful triangle: Complementarities in performance between product, process and organizational Innovation in France and the UK', *Research Policy* 44 (1), 217–32.

R. Baptista (2001), 'Geographical clusters and innovation diffusion', *Technological Forecasting and Social Change* 66 (1), 31–46.

L. Barak, E. Muller, and R. Peres (2009), 'The diffusion of services', *Journal of Marketing Research* 46 (2), 163–75.

E. Bartoloni and M. Baussola (2015), 'Persistent product innovation and market-oriented behaviour: The impact on firms' performance' (Working Paper n.105/March 2015), in *Quaderni del Dipartimento di Scienze Economiche e Sociali.* Università Cattolica del Sacro Cuore, Piacenza.

F. M. Bass (1969), 'A new product growth model for consumer durables', *Management Science* 15, 215–27.

G. Battisti, A. Canepa, and P. Stoneman (2009), 'e-Business usage across and within firms in the UK: Profitability, externalities and policy', *Research Policy* 38 (1), 133–43.

G. Battisti, M. G. Colombo, and L. Rabbiosi (2005), 'Complementarity effects in the simultaneous diffusion of technological and organisational innovations', *Industrial and Corporate Change* 24 (2), 345–82.

G. Battisti, H. Hollenstein, P. Stoneman, and M. Woerter (2007), 'Inter and intra firm diffusion of ICT in the UK and Switzerland: An internationally comparative study based on firm level data', *Economics of Innovation and New Technology* 16 (8), 669–87.

G. Battisti and A. Iona (2009), 'The intra-firm diffusion of complementary innovations: Evidence from the adoption of management practices by British establishments', *Research Policy* 38 (8), 1326–39.

G. Battisti, A. Mourani, and P. Stoneman (2010), 'Causality and a firm-level innovation scoreboard', *Economics of Innovation and New Technology* 19 (1), 7–26.

G. Battisti and P. Stoneman (2000), 'The role of regulation, fiscal incentives and changes in tastes in the diffusion of unleaded petrol in the UK', *Oxford Economic Papers* 52 (2), 326–56.

G. Battisti and P. Stoneman (2003), 'Inter- and intra-firm effects in the diffusion of new process technology', *Research Policy* 32 (9), 1641–55.

G. Battisti and P. Stoneman (2005), 'The intra-firm diffusion of new technology', *The International Journal of Industrial Organisation* 23 (1–2), 1–22.

G. Battisti and P. Stoneman (2010), 'How innovative are UK Firms? Evidence from the fourth UK Community Innovation Survey on synergies between technological and organizational innovations', *British Journal of Management* 21 (1), 187–206.

G. Battisti and P. Stoneman (2013), 'Managerial and technological innovation: UK Evidence on the sourcing, use and exploiting of new ideas'. Paper presented at the 5th ZEW/MaCCI Conference on Innovation and Patenting, Mannheim, 3–4 June.

B. Becker (2013), *The Determinants of R&D Investment: A Survey of the Empirical Research.* Loughborough: Loughborough University, School of Business and Economics. ftp://ftp.repec.org/opt/ReDIF/RePEc/lbo/lbowps/Becker_DP.pdf

S. O. Becker and P. H. Egger (2013), 'Endogenous product versus process innovation and a firm's propensity to export', *Empirical Economics* 44 (1), 329–54.

P. Beneito, P. Coscolla-Girona, M. E. Rochina-Barrachina, and A. Sanchis-Llopis (2015), 'Competitive pressure and innovation at the firm level', *Journal of Industrial Economics* 63 (3), 422–57.

G. J. Benston (1985), 'The validity of profits-structure studies with particular reference to the FTC's line of business data', *The American Economic Review* 75 (1), 37–67.

A. B. Bernard and J. Bradford Jensen (1999), 'Exceptional exporter performance: Cause, effect, or both?' *Journal of International Economics* 47 (1), 1–25.

E. R. Berndt, Z. Griliches, and N. J. Rappaport (1995), 'Econometric estimates of price indexes for personal computers in the 1990s', *Journal of Econometrics* 68 (1), 243–68.

D. Besanko and R. R. Braeutigam (2014), *Microeconomics* (5th edn). John Wiley & Sons, Inc.: Hoboken, NJ.

K. Binmore and P. Klemperer (2002), 'The biggest auction ever, the sale of the British 3G telecom licences', *Economic Journal* 112, C74–C96.

O. J. Blanchard, P. Diamond, R. E. Hall, and K. Murphy (1990), 'The cyclical behavior of the gross flows of US workers', *Brookings Papers on Economic Activity* 1990 (2), 85–155.

K. Blind (2013), *The Impact of Standardization and Standards on Innovation* (Nesta Working Paper Series 13/15). London. http://studylib.net/doc/18364231/the-impact-of-standardization-and-standards-on-innovation

B. A. Blonigen and J. Piger (2014), 'Determinants of foreign direct investment', *Canadian Journal of Economics* 47 (3), 775–812.

N. Bloom, N., M. Schankerman, and J. Van Reenen (2013), 'Identifying technology spillovers and product market rivalry', *Econometrica* 81 (4), 1347–93.

N. Bloom and J. Van Reenen (2010), 'Why do management practices differ across firms and countries?' *Journal of Economic Perspectives* 24 (1), 203–24.

A. Bohlin, H. Gruber, and B. Koutroumpis (2010), 'Diffusion of new technology generations in mobile communications', *Information Economics and Policy* 22 (1), 51–60.

S. Bond, J. A. Elston, J. Mairesse, and B. Mulkay (2003), 'Financial factors and investment in Belgium, France, Germany and the United Kingdom: A comparison using company panel data', *The Review of Economics and Statistics* 85, 153–65.

S. Bond, H. Harhoff, and J. Van Reenen (1999), 'Investment, R&D and financial constraints in Britain and Germany' (Institute for Fiscal Studies Working Paper No. 99/5). Centre for Economic Performance. http://cep.lse.ac.uk/pubs/download/dp0595.pdf

A. Borin (2008), 'Trade and quality differentiation among heterogeneous firms', Mimeo. ondazionemasi.it/public/masi/files/BORIN.pdf.

M. J. Boskin, E. L. Dulberger, R. J. Gordon, Z. Griliches, and D. W. Jorgenson (1996), *Toward a More Accurate Measure of the Cost of Living: Final Report to the Senate Finance Committee from the Advisory Commission to Study the Consumer Price Index*. Washington DC: US Government Printing Office.

R. Brealey, S. Myers, and F. Allen (2014), *Principles of Corporate Finance* (12th edn). New York: McGraw-Hill Education.

T. F. Bresnahan and R. J. Gordon (1997), 'Introduction', in T. F. Bresnahan and R. J. Gordon (eds.), *The Economics of New Goods: National Bureau of Economic Research Studies in Income and Wealth*. Chicago, IL: University of Chicago Press, 1–26.

T. Bresnahan and M. Trajtenberg (1995), 'General purpose technologies "engines of growth"?' *Journal of Econometrics* 65 (1), 83–108.

C. Broda and D. E. Weinstein (2010), 'Product creation and destruction: Evidence and price implications', *American Economic Review* 100 (3): 691–723.

E. Brouwer and A. Kleinknecht (1996), 'Firm size, small business presence and sales of innovative products: A micro-econometric analysis', *Small Business Economics* 8 (3), 189–201.

M. G. Brown and R. A. Svenson (1998), 'RTM classic: Measuring R&D productivity', *Research-Technology Management* 41 (6), 30–5.

S. L. Brown and K. M. Eisenhardt (1995), 'Product development: Past research, present findings, and future directions', *Academy of Management Review* 20 (2), 343–78.

P. Bustos (2010), 'Trade liberalization, exports, and technology upgrading: Evidence on the impact of MERCOSUR on Argentinian firms', *American Economic Review* 101 (1), 304–40.

R. D. Buzzell and F. D. Wiersema (1981a), 'Successful share-building strategies', *Harvard Business Review* 59 (1), 135–44.

R. D. Buzzell and F. D. Wiersema (1981b), 'Modelling changes in market share: A cross sectional analysis', *Strategic Management Journal* 2 (1), 27–42.

E. R. Cadotte, R. B. Woodruff, and R. L. Jenkins (1987), 'Expectations and norms in models of consumer satisfaction', *Journal of Marketing Research* 24 (3), 305–14.

A. Caldera (2010), 'Innovation and exporting: Evidence from Spanish manufacturing firms', *Review of World Economics/Weltwirtschaftliches Archiv* 146 (4), 657–89.

D. Camus (ed.) (2016), *The ONS Productivity Handbook: A Statistical Overview and Guide.* Office for National Statistics, UK. https://www.ons.gov.uk/economy/economicoutputandproductivity/productivitymeasures/methodologies/productivityhandbook#input-measures-labour-and-capital

A. Canepa and P. Stoneman (2008), 'Financial constraints to innovation in the UK: Evidence from CIS2 and CIS3', *Oxford Economic Papers* 60 (4), 711–30.

A. Cassiman, E. Golovko, and E. Martinez-Ros (2010), 'Innovation, exports and productivity', *International Journal of Industrial Organization* 28 (4), 372–6.

B. Cassiman and G. Valentini (2016), 'Open innovation: Are inbound and outbound knowledge flows really complementary?' *Strategic Management Journal* 37: 1034–46.

G. Castellion and S. K. Markham (2013), 'Perspective: New product failure rates: Influence of *argumentum ad populum* and self-interest', *Journal of Product Innovation Management* 30, 976–9. doi: 10.1111/j.1540-5885.2012.01009.

E. Cefis and M. Ciccarelli (2005), 'Profit differentials and innovation', *Economics of Innovation and New Technology* 14 (1–2), 43–61.

D. Chandrasekaran and G. J. Tellis (2007), 'A critical review of marketing research on diffusion of new products', in N. H. Malhotra (ed.), *Review of Marketing Research*, vol. 3. Armonk, NY: Emerald Group Publishing, 39–80.

P. K. Chaney, T. M. Devinney, and R. S. Winer (1991), 'The impact of new product introductions on the market value of firms', *Journal of Business* 64 (4), 573–610.

T. Z. Chang and Su-Jane Chen (1998), 'Market orientation, service quality and business profitability: A conceptual model and empirical evidence', *Journal of Services Marketing* 12 (4), 246–64.

S.-S. Chen, K. W. Ho, K. H. Ik, and Cheng-few Lee (2002), 'How does strategic competition affect firm values? A study of new product announcements', *Financial Management* 3, 67–84.

H. W. Chesbrough (2003), 'The era of open innovation', *MIT Sloan Management Review*, 44 (3), 35–41.

M. Cincera and O. Galgau (2005), *Impact of Market Entry and Exit on EU Productivity and Growth Performance* (Economic Papers, No. 222). European Economy, European Commission,

Directorate General for Economic and Financial Affairs. http://ec.europa.eu/economy_finance/publications/publication712_en.pdf

G. Clark (1984), *Innovation Diffusion: Contemporary Geographical Approaches* (Concepts and Techniques in Modern Geography 40). Geo Books: Norwich.

W. Cohen (1995), 'Empirical studies of innovative activity', in P. Stoneman (ed.), *Handbook of the Economics of Innovation and Technological Change.* Oxford: Blackwell, 182–264.

W. Cohen (2010), 'Fifty years of empirical studies of innovative activity and performance', in B. Hall and N. Rosenberg (eds.), *Handbook of the Economics of Innovation*, vol. 1. Amsterdam: Elsevier, 129–213.

D. A. Comin and B. Hobijn (2004), 'Cross-country technology adoption: Making theory face the facts', *Journal of Monetary Economics* 51 (1), 39–83.

D. A. Comin and B. Hobijn (2009), *The CHAT Dataset* (Working Paper 15319). Harvard Business School and NBER. http://www.hbs.edu/faculty/Publication%20Files/10-035.pdf

D. A. Comin and B. Hobijn (2010), 'An exploration of technology diffusion', *American Economic Review* 100 (5), 2031–59.

D. A. Comin, B. Hobijn, and E. Rovito (2006), *Five Facts You Need to Know About Technology Diffusion* (NBER Working Papers 11928). NBER. https://www.researchgate.net/publication/5186709_Five_Facts_You_Need_to_Know_About_Technology_Diffusion

D. A. Comin, B. Hobijn, and E. Rovito (2008a), 'A new approach to measuring technology with an application to the shape of the diffusion curves', *Journal of Technology Transfer* 33 (2), 187–207.

D. A. Comin, B. Hobijn, and E. Rovito (2008b), 'Technology usage lags', *Journal of Economic Growth* 13 (4), 237–56.

D. A. Comin and M. Mestieri (2010), *The Intensive Margin of Technology Adoption* (Working Paper No. 11 026). Harvard Business School, BGIE Unit. http://economics.mit.edu/files/6224

D. A. Comin and M. Mestieri (2013), *Technology Diffusion: Measurement, Causes and Consequences* (NBER Working Paper 19052). NBER. http://www.dartmouth.edu/~dcomin/files/chapter_v8.pdf

D. A. Comin and M. Mestieri (2014), 'Technology diffusion: Measurement, causes and consequences', in P. Aghion and S. Durlauf (eds.), *Handbook of Economic Growth*, vol. 2. North Holland: Elsevier, 565–622.

R. A. Connolly and S. Schwartz (1985), 'The intertemporal behavior of economic profits', *International Journal of Industrial Organization* 3 (4), 379–400.

R. Cooper, S. Edgett, and E. Kleinschmidt (2001), 'Portfolio management for new product development: Results of an industry practices study', *R&D Management* 31 (4), 361–80.

M. Corsino, G. Espa, and R. Micciolo (2011), 'R&D, firm size and incremental product innovation', *Economics of Innovation and New Technology* 20 (5), 423–43.

B. P. Cozzarin (2016), Impact of organizational innovation on product and process innovation', *Economics of Innovation and New Technology* 18 (2), 1–13.

D. Crass (2014), *The impact of brand use on innovation performance: Empirical results for Germany* (Discussion Paper 14-119). ZEW-Centre for European Economic Research. http://ftp.zew.de/pub/zew-docs/dp/dp14119.pdf

B. Crepon, E. Duguet, and J. Mairesse (1998), 'Research, innovation and productivity: An econometric analysis at the firm level', *Economics of Innovation and New Technology* 7 (2), 115–58.

M. Crozet, K. Head, and T. Mayer (2012), 'Quality sorting and trade: Firm-level evidence for French wine', *Review of Economic Studies* 79 (2), 609–44.

M. A. Cusumano, Y. Mylonadis, and R. S. Rosenbloom (1992), 'Strategic manoeuvring and mass market dynamics: VHS over Beta', *Business History Review* 66 (1), 51–94.

S. G. Dacko, P. Stoneman, and Z. Kastrinaki (2015), 'New product introduction: Follower firm timing behaviour', *Economics of Innovation and New Technology* 24 (8), 829–53.

J. P. Damijan, C. Kostevc, and S. Polanec (2010), 'From innovation to exporting or vice versa?' *World Economy* 33 (3), 374–98.

A. J. d'Aspremont, J. Gabszewicz, and J.-F. Thisse (1979), 'On Hotelling's stability in competition', *Econometrica* 47 (5), 1145–50.

P. Dasgupta and P. A. David (1994), 'Toward a new economics of science', *Policy Research* 23, 487–521.

P. Dasgupta and J. Stiglitz (1980), 'Industrial structure and the nature of innovative activity', *Economic Journal* 90, 266–93.

P. Dasgupta and P. Stoneman (eds.) (1987), *Economic Policy and Technological Performance.* Cambridge: Cambridge University Press.

P. David and S. Greenstein (1990), 'The economics of compatibility standards: An introduction to recent research', *Economics of Innovation and New Technology* 1, 3–41.

P. David, P. Hall, and A. Toole (2000), 'Is public R&D a complement or substitute for private R&D? A review of the econometric evidence', *Research Policy* 29 (4–5), 497–529.

T. Davila, M. Epstein, and R. Shelton (2012), *Making Innovation Work: How to Manage It, Measure It, and Profit from It.* Upper Saddle River, NJ: FT Press.

C. R. Davis (2002), 'Calculated risk: A framework for evaluating product development', *MIT Sloan Management Review* 43 (4), 71–7.

S. J. Davis and J. Haltiwanger (1996), 'On the driving forces behind cyclical movement, in employment and job reallocation', *American Economic Review* 89 (5), 1234–58.

G. S. Day (1994), 'The capabilities of market-driven organizations', *Journal of Marketing* 58 (4), 37–52.

J. P. J. de Jong and P. A. M. Vermeulen (2006), 'Determinants of product innovation in small firms', *International Small Business Journal* 24 (6), 587–609.

A. Deaton (1998), 'Getting prices right: What should be done?' *Journal of Economic Perspectives* 12 (1), 37–46.

A. Dechezleprêtre, E. Einiö, R. Martin, K-T. Nguyen, and J. Van Reenen (2016), *Do Tax Incentives for Research Increase Firm Innovation?* (CEPR Discussion Paper No 1413). Centre for Economic Performance, London School of Economics. http://cep.lse.ac.uk/pubs/download/dp1413.pdf

Defence Committee (2004), *Defence Procurement.* London: House of Commons, Stationery Office. https://publications.parliament.uk/pa/cm200304/cmselect/cmdfence/572/572.pdf

Department for Business, Energy and Industrial Strategy [DBEIS] (2016), *UK Innovation Survey 2015: Statistical Annex and Interactive Report.* https://www.gov.uk/government/statistics/uk-innovation-survey-2015-statistical-annex-and-interactive-report

Department for Business Innovation and Skills [DBIS] (2013), *First Findings from the UK Innovation Survey, 2011 (Revised).* Science and Innovation Analysis. https://www.gov.uk/government/uploads/system/uploads/attachment_data/file/200078/12-P106A-UKIS_2011First_findings_Apr13.pdf

Department for Business Innovation and Skills [DBIS] (2015), *UK Innovation Survey 2012 to 2014: Statistical Annex.* https://www.gov.uk/government/statistics/uk-innovation-survey-2015-statistical-annex-and-interactive-report.

Department for Business Innovation and Skills [DBIS] (2016a), *Headline Findings from the UK Innovation Survey 2015: Innovation Analysis*, July, https://www.gov.uk/government/uploads/system/uploads/attachment_data/file/506953/bis-16-134-uk-innovation-survey-2015.pdf

Department for Business Innovation and Skills [DBIS] (2016b), *The UK Innovation Survey 2015.* The Main Report, July, https://www.gov.uk/government/uploads/system/uploads/attachment_data/file/536491/UKIS_2015_Main__report_Final_v.pdf

Department of Culture Media and Sport (2006), *Creative Industries, Economic Estimates, Statistical Bulletin.* September. London.

R. Deshpande and F. E. Webster Jr. (1989), 'Organizational culture and marketing: Defining the research agenda', *Journal of Marketing* 53 (1), 3–15.

M. D. Dickerson and J. W. Gentry (1983), 'Characteristics of adopters and non-adopters of home computers', *Journal of Consumer Research* 10, 225–35.

W. E. Diewert (2002), Harmonized Indexes of Consumer Prices: Their Conceptual Foundations (Working Papers Series 130). European Central Bank. https://www.ecb.europa.eu/pub/pdf/scpwps/ecbwp130.pdf?7f9fcb71c9a79e446251a79273c9ba60

C. Dimos (2016), 'The effectiveness of R&D subsidies: A meta regression analysis of the evaluation literature', *Research Policy* 45 (4), 797–815.

A. Dixit and R. Pindyck (1994), *Investment under Uncertainty.* Princeton, NJ: Princeton University Press.

A. K. Dixit and J. E. Stiglitz (1977), 'Monopolistic competition and optimum product diversity', *American Economic Review* 67, 297–308.

J. Doran (2012), 'Are differing forms of innovation complements or substitutes?' *European Journal of Innovation Management* 15 (3), 351–71.

G. Dosi (1982), 'Technological paradigms and technological trajectories: A suggested interpretation of the determinants and directions of technical change', *Research Policy* 11 (3), 147–62.

M. du Plessis (2007), 'The role of knowledge management in innovation', *Journal of Knowledge Management* 11 (4), 20–9.

C. Duhigg and K. Bradsher (2012), 'How the US lost out on iPhone work', *New York Times*, 21 January. http://www.nytimes.com/2012/01/22/business/apple-america-and-a-squeezed-middle-class.html

J. Eaton and S. Kortum (1999), 'International patenting and technology diffusion: Theory and measurement', *International Economic Review* 40 (3), 537–70.

J. Eaton and S. Kortum (2001), 'Trade in capital goods', *European Economic Review* 45 (7), 1195–235.

N. Economides (1989), 'Desirability of compatibility in the absence of network externalities', *American Economic Review* 78 (1), 108–21.

C. Edquist, L. Hommen, and M. McKelvey (2001), *Innovation and Employment: Process Versus Product Innovation.* Cheltenham: Edward Elgar.

J. Edwards, J. A. Kay, C. Mayer, and J. Edwards (1987), *The Economic Analysis of Accounting Profitability.* Oxford: Oxford University Press.

C. Ennen and A. Richter (2010), 'The whole is more than the sum of its parts—or is it? A review of the empirical literature on complementarities in organizations', *Journal of Management* 36 (1), 207–33.

S. K. Ethiraj and D. H. Zhu (2008), 'Performance effects of imitative entry', *Strategic Management Journal* 29 (8), 797–817.

Eurostat Task Force (1999), 'Volume Measures for Computers and Software', *Report of the Eurostat Task Force on Volume Measures for Computers and Software, June.* Luxembourg: Eurostat.

Eurostat (2001), *Handbook on Price and Volume Measures in National Accounts*, Luxembourg: Office for Official Publications of the European Communities.

Eurostat (2016), *Obstacles to Innovation Considered Highly Important by Innovative and Non-Innovative Enterprises*, European Commission Database, March. Found via Eurostat data explorer at appso.eurostat.ec.europa.eu/nui/show.do?dataset=inncis8obst&lang=en.

C. F. Fama and K. R. French (1993), 'Common risk factors in the returns on stocks and bonds', *Journal of Financial Economics* 33 (1), 3–56.

D. Fantino (2008), *R&D and Market Structure in a Horizontal Differentiation Framework* (Economic Working Paper 658). Bank of Italy, Economic Research and International Relations Area.

G. Feder and G. T. O'Mara (1982), 'Farm size and the diffusion of green revolution technology, on information and innovation diffusion: A Bayesian approach', *American Journal of Agricultural Economics* 64 (1), 145–7.

The Federation of European Publishers (2012), *European Book Publishing Statistics 2011.* http://www.fep-fee.eu/FEP-Statistics-for-the-year-2011

R. C. Feenstra (2015), *Advanced International Trade: Theory and Evidence.* Princeton, NJ: Princeton University Press.

R. C. Feenstra and G. H. Hanson (1996), 'Foreign investment, outsourcing and relative wages', in R. C. Feenstra (ed.), *Political Economy of Trade Policy: Essays in Honor of Jagdish Bhagwati.* Cambridge, MA: MIT Press, 89–128.

C. Fehle, S. M. Fournier, T. J. Madden, and D. G. Shrider (2008), 'Brand value and asset pricing', *Quarterly Journal of Finance and Accounting* 47 (1), 3–26.

S. Feigl and K. Menrad (2008), *Innovation Activities in the Food Industry in Selected European Countries.* Straubing: University of Applied Sciences of Weihenstephan, Science Center. http://www.wz-straubing.de/fachhochschule-weihenstephan/images/cms/innovations%20food%20industry.pdf

M. P. Feldman and D. F. Kogler (2010), 'Stylized facts in the geography of innovation', in B. Hall and N. Rosenberg (eds.), *Handbook of the Economics of Innovation*, vol. 1. Amsterdam: Elsevier, 381–410.

R. D-S. Ferreira and J.-F. Thisse (1996), 'Horizontal and vertical differentiation: The Launhardt model', *International Journal of Industrial Organization* 14 (4), 485–506.

A. Filippetti (2011), 'Innovation modes and design as a source of innovation: A firm-level analysis', *European Journal of Innovation Management* 14 (1), 5–26.

F. M. Fisher and J. J. McGowan (1983), 'On the misuse of accounting rates of return to infer monopoly profits', *American Economic Review* 73 (1), 82–97.

M. T. Flaherty (1981), *Market Share, Technology Leadership, and Competition in International Semiconductor Markets.* Cambridge, MA: Division of Research, Graduate School of Business Administration, Harvard University.

D. Frantzen (2003), 'The causality between R&D and productivity in manufacturing: An international disaggregate panel data study', *International Review of Applied Economics* 17 (2), 125–46.

C. Freeman (2000), *The Economics of Industrial Innovation*. Abingdon: Taylor & Francis.

M. Frenz and R. Lambert (2008), 'Technological and non-technological innovation', in *Science, Technology and Industry Outlook 2008*. Paris: OECD Publications, 235–8.

M. Frenz and R. Lambert (2009), 'Exploring non-technological and mixed modes of innovation across countries', in *Innovation in Firms: A Microeconomic Perspective*. Paris: OECD publications, 69–110., http://www.keepeek.com/oecd/media/science-and-technology/innovation-in-firms_9789264056213-en#.WZ_9uLhGTBg#page69

M. Fritsch and M. Meschede (2001), 'Product innovation, process innovation, and size', *Review of Industrial Organization* 19 (3), 335–50.

D. Fudenberg and J. Tirole (1985), 'Pre-emption and rent equalization in the adoption of new technology', *Review of Economic Studies* 52 (3), 383–401.

J. Gabszewicz and J. Thisse (1979), 'Price competition, quality and income disparities', *Journal of Economic Theory* 20 (3), 340–59.

J. Gabszewicz and O. Tarola (2012), 'Product innovation and firms' ownership', *Economics of Innovation and New Technology* 21 (4), 323–43.

P. Ganotakis and J. H. Love (2011), 'R&D, product innovation, and exporting: evidence from UK new technology based firms', *Oxford Economic Papers* 63 (2), 279–306.

Z. Gao and C. Tisdell (2004), *Television Production: Its Changing Global Location, the Product Cycle and China* (Working Paper 26). Brisbane: School of Economics, University of Queensland.

H. Gatignon and J. M. Xuereb (1997), 'Strategic orientation of the firm and new product performance', *Journal of Marketing Research* 34, 77–90.

P. Geroski (1990), 'Innovation technological opportunity and market structure', *Oxford Economic Papers* 42 (3), 586–602. doi: https://doi.org/10.1093/oxfordjournals.oep.a041965

P. Geroski (1995), 'Markets for technology: Knowledge, innovation and appropriability', in P. Stoneman (ed.), *Handbook of the Economics of Innovation and Technological Change*. Oxford: Blackwell, 90–131.

P. Geroski (2003), *The Evolution of New Markets*. Oxford: Oxford University Press.

P. Geroski, S. Machin, and J. Van Reenen (1993), 'The profitability of innovating firms', *RAND Journal of Economics* 24 (2), 198–211.

S. Glaister (1974), 'Advertising policy and returns to scale in markets where information is passed between individuals', *Economica* 41, 139–46.

M. Gort and S. Klepper (1982), 'Time paths in the diffusion of product innovations', *Economic Journal* 92 (367), 630–53.

G. E. Greenley (1995), 'Market orientation and company performance: Empirical evidence from UK Companies', *British Journal of Management* 6 (1), 1–13.

S. Greenstein and G. Ramey (1998), 'Market structure, innovation and vertical product differentiation', *International Journal of Industrial Organisation*, 16 (3), 285–311.

Z. Griliches (1957), 'Hybrid corn: An exploration in the economics of technological change', *Econometrica* 25, 501–22.

Z. Griliches (1990), 'Patent statistics as economic indicators: a survey', *Journal of Economic Literature* 28, 1661–707.

G. M. Grossman and E. Helpman (1991), *Innovation and Growth in the Global Economy*. Cambridge, MA: MIT Press.

B. Hall (1993), 'The stock market's valuation of R&D investment during the 1980s', *American Economic Review* 83 (2), 259–64.

B. Hall (1999), *Innovation and Market Value* (Working Paper 6984). NBER. http://www.nber.org/papers/w6984.pdf

B. Hall (2002), 'The financing of research and development', *Oxford Review of Economic Policy* 18, 35–51.

B. Hall (2011), *Innovation and Productivity* (Working Paper 17178). NBER. http://www20.iadb.org/intal/catalogo/pe/2011/08562.pdf

B. Hall, P. M-P. Castello, S. Montresor, and A. Vezzani (2016), 'Financing constraints, R&D investments and innovative performances: New empirical evidence at the firm level for Europe', *Economics of Innovation and New Technology* 25 (3), 183–96.

B. Hall and D. Harhoff (2012), 'Recent research on the economics of patents', *Annual Review of Economics, Annual Reviews* 4 (1), 541–65.

B. Hall, C. Helmers, M. Rogers, and V. Sena (2013), 'The importance (or not) of patents to UK firms', *Oxford Economic Papers* 65 (3), 603–29.

B. Hall and J. Lerner (2010), 'The financing of R&D and innovation', in B. Hall and N. Rosenberg (eds.), *Handbook of the Economics of Innovation*, vol. 1. Amsterdam: Elsevier, 609–39.

B. Hall, J. Mairesse, and P. Mohnen (2010), 'Measuring the returns to R&D', in B. Hall and N. Rosenberg (eds.), *Handbook of the Economics of Innovation*, vol. 2. Amsterdam: Elsevier, 1033–82.

B. Hall and R. Oriani (2006), 'Does the market value R&D investment by European firms? Evidence from a panel of manufacturing firms in France, Germany, and Italy', *International Journal of Industrial Organization* 24 (5), 971–93.

W. K. Hall (1980), 'Survival Strategies in a Hostile Environment', *Harvard Business Review* 58 (5), 75–85.

J. K. Han, N. Kim, and R. K. Srivastava (1998), 'Market orientation and organizational performance: is innovation a missing link?' *Journal of Marketing* 62 (4), 30–45.

L. C. Harris (2001), 'Market orientation and performance: Objective and subjective empirical evidence from UK companies', *Journal of Management Studies* 38 (1), 17–43.

J. Hauser, G. J. Tellis, and A. Griffin (2006), 'Research on innovation: A review and agenda for marketing science', *Marketing Science* 25 (6), 687–717.

J. A. Hausman (1996), 'Valuation of new goods under perfect and imperfect competition', in T. F. Bresnahan and R. J. Gordon (eds.), *Economics of New Goods: National Bureau of Economic Research Studies in Income and Wealth.* Chicago, IL: University of Chicago Press, 207–48.

J. A. Hausman (1999), 'Cellular telephone, new products, and the CPI', *Journal of Business & Economic Statistics* 17 (2), 188–94.

J. A. Hausman (2003), 'Sources of bias and solutions to bias in the consumer price index', *The Journal of Economic Perspectives* 17 (1), 23–44.

G. Hawawini, V. Subramanian, and P. Verdin (2003), 'Is performance driven by industry or firm specific factors? A new look at the evidence', *Strategic Management Journal* 24 (1), 1–16.

S. A. Heffernan, Xialon Fu, and Xiaoqing Fu (2013), 'Financial innovation in the UK', *Applied Economics* 45 (24), 3400–11.

K. B. Hendricks and V. R. Singhal (1997), 'Delays in new product introductions and the market value of the firm: The consequences of being late to the market', *Management Science* 43 (4), 422–36.

I. Herguera and S. Lutz (2010), 'The effect of subsidies to product innovation on international competition', *Economics of Innovation and New Technology* 12 (5), 465–80.

H. Hotelling (1929), 'Stability in competition', *Economic Journal* 39 (153), 41–57.

R. G. Hubbard (1998), 'Capital market imperfections and investment', *Journal of Economic Literature* 36, 193–225.

R. F. Hurley and G. T. M. Hult (1998), 'Innovation, market orientation, and organizational learning: An integration and empirical examination', *Journal of Marketing* 62, 42–54.

C. Ichniowski, K. Shaw, and G. Prennushi (1997), 'The effects of human resource management practices on productivity: A study of steel finishing lines', *American Economic Review* 87 (3), 291–313.

C. Imbriani, P. Morone, and F. Renna (2015), 'Innovation and exporting: Does quality matter?', *International Trade Journal* 29 (4), 273–90.

Intellectual Property Office (2014), *Exceptions to Copyright: An Overview*. Cardiff: IPO.

International Federation of the Phonographic Industry [IFPI] (2016), *Global Music Report 2016*. http://ifpi.org/news/IFPI-GLOBAL-MUSIC-REPORT-2016

N. Ireland and P. Stoneman (1986), 'Technological diffusion, expectations and welfare', *Oxford Economic Papers* 38, 283–304.

A. Jaffe, M. Trajtenberg, and R. Henderson (1993), 'Geographic localization of knowledge spill overs as evidenced by patent citations', *Quarterly Journal of Economics* 108 (3), 577–98.

S. S. Jang and K. M. Park (2012), *Differentiation Strategy Revisited: The Performance Effects of Vertical and Horizontal Differentiation*. Yonsei Business Research Institute, Korea. https://ybri.yonsei.ac.kr/downloadfile.asp?wpid=25&mid=m02%5F01 . . .

J. Jaumandreu and J. Mairesse (2010), *Innovation and Welfare: Results from Joint Estimation of Production and Demand Functions* (Working Paper 16221). NBER. http://people.bu.edu/jordij/papers/w16221.pdf

J. Jaumandreu and J. Mairesse (2016), 'Disentangling the effects of process and product innovations on cost and demand', *Economics of Innovation and New Technology* 26, 1–18. doi: 10.1080/10438599.2016.1205276.

R. Jensen (1988), 'Information costs and innovation adoption policies', *Management Science INFORMS* 34, 230–9.

J. Jewkes, D. Sawers, and R. Stillerman (1969), *The Sources of Invention*. New York: W. W. Norton.

B. Jing (2006), 'On the profitability of firms in a differentiated industry', *Marketing Science* 25 (3), 248–59.

D. S. Johnson, S. B. Reed, and K. J. Stewart (2006), 'Price measurement in the United States: A decade after the Boskin Report'. US Bureau of Labor Statistics, Monthly Labour Review. https://www.bls.gov/opub/mlr/2006/05/art2full.pdf

C. Jones and J. Williams (2000), 'Too much of a good thing? The economics of investment in R&D', *Journal of Economic Growth* 5, 66–85.

U. Kaiser (2007), 'A microeconomic note on product innovation and product innovation advertising', *Economics of Innovation and New Technology* 15 (7), 573–82.

K. Kaivanto (2006), 'Premise and practice of UK launch aid', *Journal of World Trade* 40 (3), 495–525.

K. Kaivanto and P. Stoneman (2004), *Risk Shifting Technology Policy and Sales Contingent Claims: When Is Launch Aid to the Aerospace Industry a Subsidy* (CEPR Discussion Paper No 4798). London: Centre for Economic Policy and Research.

K. Kaivanto and P. Stoneman (2007), 'Public provision of sales contingent claims backed finance to SMEs: a policy alternative', *Research Policy* 36 (5), 637–51.

A. Kandybin (2009), 'Which innovation efforts will pay?' *MIT Sloan Management Review* 51 (1), 53.

M. Karshenas and P. Stoneman (1993), 'Rank, stock, order and epidemic effects in the diffusion of new process technologies: An empirical model', *RAND Journal of Economics* 24 (4), 503–28.

I. C. Kerssens-van Drongelen and A. Cook (1997), 'Design principles for the development of measurement systems for research and development processes', *R&D Management* 27 (4), 345–57.

S. Klepper (1996), 'Entry, exit, growth and innovation over the product life cycle', *American Economic Review* 86 (3), 562–83.

S. Klepper and E. Grady (1990), 'The evolution of new industries and the determinants of market structure', *RAND Journal of Economics* 21 (1), 27–44.

C. Köhler, P. Laredo, and C. Rammer (2012), *The Impact and Effectiveness Of Fiscal Incentives for R&D* (Nesta Working Paper 12/01). London: Nesta.

V. Krishnan, K. T. Ulrich (2001), 'Product development decisions: A review of the literature', *Management Science* 47 (1), 1–21.

T. D. Kuczmarski (2000), 'Measuring your return on innovation', *Marketing Management* 9 (1), 24–32.

L. Lambertini and R. Orsini (2000), 'Process and product innovation in a vertically differentiated monopoly', *Economics Letters* 68 (3), 333–7.

M. Lambkin (1988), 'Order of entry and performance in new markets', *Strategic Management Journal* 9 (S1), 127–40.

F. Lamperti, R. Mavilia, and M. Giometti (2016), 'Persistence of innovation and knowledge flows in Africa: An empirical investigation', *Innovation and Development* 6 (2), 235–57.

K. J. Lancaster (1966), 'A new approach to consumer theory', *Journal of Political Economy* 74 (2), 132–57.

K. J. Lancaster (1990), 'The economics of product variety: A survey', *Marketing Science* 9 (3), 189–206.

G. Lanzolla, J. Gomez, and J. P. Maicas (2010), 'Order of market entry, market and technological evolution and firm competitive performance'. Paper presented at the DRUID Conference, Imperial College, London, 16–18 June.

R. Levin, A. Klevorick, R. Nelson, and S. Winter (1987), 'Appropriating the Returns from Industrial R&D', *Brookings Papers on Economic Activity* 18 (3), 783–832.

S. Lhuillery (2014), 'Marketing and persistent innovation success', *Economics of Innovation and New Technology* 23 (5–6), 517–43.

B. Libai, E. Muller, and R. Peres (2009), 'The diffusion of services', *Journal of Marketing Research* 46 (2), 163–75.

M. B. Lieberman and D. B. Montgomery (1988), 'First-mover advantages', *Strategic Management Journal* 9 (S1), 41–58.

J. Liikanen, P. Stoneman, and O. Toivanen (2004), 'Intergenerational effects in the diffusion of new technology: the case of mobile phones', *International Journal of Industrial Organization* 22 (8–9), 1137–54.

H. L. Lin and E. S. Lin (2010), 'FDI, trade and product innovation', *Southern Economic Journal* 77 (2), 434–64.

P. Lin and K. Saggi (2002), 'Product differentiation, process R&D, and the nature of market competition', *European Economic Review* 46 (1), 201–11.

E. B. Lindenberg and S. A. Ross (1981), 'Tobin's Q ratio and industrial organization', *Journal of Business* 54 (1), 1–32.

D. Lindsay (2002), *The Law and Economics of Copyright, Contract and Mass Market Licences*. Strawberry Hills, Australia: Centre for Copyright Studies Ltd. https://static-copyright-com-au.s3.amazonaws.com/uploads/2015/08/CentreCopyrightStudies_AllenGroup-TheLawandEconomicsofCopyright.pdf

H. Lofsten (2014), 'Product innovation processes and the trade-off between product innovation performance and business performance', *European Journal of Innovation Management* 17 (1), 61–84.

H. Loof and A. Heshmati (2006), 'On the relationship between innovation and performance: A sensitivity analysis', *Economics of Innovation and New Technology* 15 (4–5), 317–44.

B-Å. Lundvall and A. L. Vinding (2004), 'Product innovation and economic theory: User–producer interaction in the learning economy', in J. L. Christensen, and B. Å. Lundvall (eds.), *Product Innovation, Interactive Learning and Economic Performance*. Amsterdam: Elsevier, 101–28.

S. Lutz (1996), *Vertical Product Differentiation and Entry Deterrence* (CEPR Discussion Paper 1455). London: CEPR.

Madakom (ed.) (2001), *Innovations Report 2001*. Madakom GmbH: Köln.

V. Mahajan, E. Muller, and Y. Wind (eds.) (2000), *New-Product Diffusion Models*. New York: Springer.

E. Mansfield (1963), 'Intrafirm rates of diffusion of an innovation', *Review of Economics and Statistics* 45, 348–59.

E. Mansfield (1968), *Industrial Research and Technological Innovation: An Econometric Analysis*. New York: W. W. Norton for the Cowles Foundation.

H. M. Markowitz (1952), 'Portfolio selection', *Journal of Finance* 7 (1), 77–91. doi: 10.2307/2975974.

B. McCall (2014), *The Economics of Search*. London: Routledge.

K. T. McNamara, C. R. Weiss, and A. Wittkopp (2003), *Market Success of Premium Product Innovation: Empirical Evidence from the German Food Sector* (Working Paper FE 0306). Department of Food Economics and Consumption Studies, University of Kiel.

M. J. Melitz (2003), 'The impact of trade on intra-industry reallocations and aggregate industry productivity', *Econometrica* 71 (6), 1695–725.

P. Milgrom and J. Roberts (1990), 'The economics of modern manufacturing: Technology, strategy and organization', *American Economic Review* 80 (3), 511–28.

P. Milgrom and J. Roberts (1995), 'Complementarities and fit, strategy, structure, and organizational change in manufacturing', *Journal of Accounting and Economics* 19 (2), 179–208.

P. Milling and J. Stumpfe (2000), 'Product and process innovation a system dynamics-based analysis of the interdependencies'. Systems Dynamic Society. http://www.systemdynamics.org/conferences/2000/PDFs/milling1.pdf

A. Mina, H. Lahr, and A. Hughes (2013), 'The demand and supply of external finance for innovative firms', *Industrial and Corporate Change* 22 (4), 869–901.

L. Mlodinow and S. Hawking (2008), *A Briefer History of Time*. London: Bantam Press.

P. Mohnen and L-Hendrik Roller (2005), 'Complementarities in innovation policy', *European Economic Review* 49 (6), 1431–50.

Y. Moon (2005), 'Break free from the product life cycle', *Harvard Business Review* 83 (5): 86–94.

T. H. Moran (2012), 'Foreign direct investment', in G. Ritzer (ed.), *Wiley Blackwell Encyclopaedia of Globalization*. doi: 10.1002/9780470670590.wbeog216

E. Moretti, C. Steinwender, and J. Van Reenen (2016), 'The intellectual spoils of war? defense R&D', Econometrics Laboratory, UC Berkeley, 8 July. http://eml.berkeley.edu/~moretti/military.pdf

R. Morris (2009), *The Fundamentals of Product Design*. Lausanne: AVA Publishing.

F. Mostacci (2008), 'Gli aggiustamenti di qualità negli indici dei prezzi al consumo in Italia: Metodi, caso di studio e indicatori impliciti', *Contributi ISTAT* 3, 5–26. http://www3.istat.it/dati/pubbsci/contributi/Contributi/contr_2008/03_2008.pdf

D. C. Mueller and J. E. Tilton (1969), 'Research and development costs as a barrier to entry', *Canadian Journal of Economics/Revue Canadienne d'Économique* 2 (4), 570–9.

M. D. Mumford (2003), 'Where have we been, where are we going? Taking stock in creativity research', *Creativity Research Journal* 15, 107–20.

S. Nagaoka, K. Motohashi, and A. Goto (2010), 'Patent statistics as an innovation indicator', in B. Hall and N. Rosenberg (eds.), *Handbook of the Economics of Innovation*, vol. 2. Amsterdam: Elsevier, 1083–127.

J. C. Narver and S. F. Slater (1990), 'The effect of a market orientation on business profitability', *Journal of Marketing* 54 (4), 20–35.

National Audit Office (2004), *Improving IT procurement* (Report by the Comptroller and Auditor General, November). London: The Stationery Office.

R. R. Nelson (ed.) (1993), *National Innovation Systems: A Comparative Analysis*. Oxford: Oxford University Press.

D. Neven and J.-F. Thisse (1989), *On Quality and Variety Competition* (CORE Discussion Paper 1989020). CORE, Université Catholique de Louvain.

Y.-H. Noh and G. Moschini (2006), 'Vertical product differentiation, entry-deterrence strategies, and entry qualities', *Review of Industrial Organization* 29, 227–52.

E. M. Olson, O. C. Walker Jr., and R. W. Ruekert (1995), 'Organizing for effective new product development: The moderating role of product innovativeness', *Journal of Marketing*, 59, 48–62.

Organisation for Economic Co-Operation and Development [OECD] (2002), *The Measurement of Scientific and Technological Activities, Proposed Standard Practice for Surveys on Research and Experimental Development* (Frascati Manual). Paris: OECD.

Organisation for Economic Co-Operation and Development [OECD] (2005), *Guidelines For Collecting and Interpreting Innovation Data* (Oslo Manual, 3rd edn). Paris: OECD/Eurostat.

PA Consulting Limited (2014), *Innovation for Peak Performance: Levers to Improve Your Innovation Capabilities towards R&D Excellence*. PA Knowledge Limited: London.

E. T. Penrose (1959), *The Theory of the Growth of the Firm*. New York: Sharpe.

R. Peres, V. Mahajan, and E. Muller (2009), 'Innovation diffusion and new product growth models: A critical review and research directions', *International Journal of Research in Marketing* 27 (2), 91–106.

G. Peri (2005), 'Determinants of knowledge flows and their effect on innovation', *Review of Economics and Statistics* 87 (2), 308–22.

B. Peters, M. J. Roberts, and V. A. Vuong (2016), *Dynamic R&D Choice and the Impact of the Firm's Financial Strength* (Working Paper 22035). NBER. https://www.uts.edu.au/sites/default/files/cdm_finance_final.pdf

A. Petrin and F. Warzynski (2012), *The Impact of Research and Development on Quality, Productivity and Welfare*. Mimeo, University of Minnesota. https://docs.google.com/a/umn.edu/viewer?a=v&pid=sites&srcid=dW1uLmVkdXxhbWlsLXBldHJpbnxneDo0NjIxYTI4MDE2N2FkNzY1

L. W. Phillips, D. R. Chang, and R. D. Buzzell (1983), 'Product quality, cost position and business performance: A test of some key hypotheses', *Journal of Marketing* 47 (2), 26–43.

G. P. Pisano (2015), 'You need an innovation strategy', *Harvard Business Review* 93 (6): 44–54.

R. Pitkethly (2007), *UK Intellectual Property Awareness Survey, 2006*. Intellectual Property Office. http://webarchive.nationalarchives.gov.uk/20140603103830/http://www.ipo.gov.uk/ipsurvey2010.pdf

M. E. Porter (1980), *Competitive Strategy: Techniques for Analyzing Industries and Competitor*. New York: Free Press.

L. Prosser (ed.) (2009), *UK Standard Industrial Classification of Economic Activities 2007 (SIC 2007): Structure and explanatory notes*. Bristol: Office for National Statistics.

A. M. Pulkki-Brännström and P. Stoneman (2013), 'On the patterns and determinants of the global diffusion of new technology', *Research Policy* 42 (10), 1768–79.

H. Rao and R. Drazin (2002), 'Overcoming resource constraints on product innovation by recruiting talent from rivals: A study of the mutual funds industry 1986–1994', *Academy of Management Journal* 45 (3), 491–507.

Recording Industry Association of Japan (2016), *RIAJ Yearbook, Statistics and Trends 2016*. RIAJ. http://www.riaj.or.jp/f/pdf/issue/industry/RIAJ2016E.pdf

J. F. Reinganum (1981), 'Market structure and the diffusion of new technology', *Bell Journal of Economics* 12, 618–24.

M. Reisinger (2004), *Vertical Product Differentiation, Market Entry, and Welfare* (Discussion Papers in Economics 479). Department of Economics, University of Munich.

W. D. Reitsperger, S. J. Daniel, S. B. Tallman, and W. G. Chismar (1993), 'Product quality and cost leadership: Compatible strategies?' *MIR: Management International Review* 33 (1), 7–21.

R. Rigobon and D. Rodrik (2005), 'Rule of law, democracy, openness, and income: Estimating the interrelationships', *Economics of Transition* 13 (3), 533–64.

P. W. Roberts (1999), 'Product innovation, product-market competition and persistent profitability in the US pharmaceutical industry', *Strategic Management Journal* 20 (7), 655–70.

P. W. Roberts (2001), 'Innovation and firm level persistent profitability: A Schumpeterian framework', *Managerial and Decision Economics* 22 (4–5), 239–50.

S. Robson and L. Ortmans (2006), 'First Findings from the UK Innovation Survey 2005', *Economic Trends* 628, 58–64.

E. M. Rogers (2003), *Diffusion of Innovations* (5th edn.). New York: Free Press.

D. Rombach (2014), *Fraunhofer as a Success Model for Applied Research and Technology Transfer*. Luxembourg, 19 February. people.svv.lu/lectures/20140219-Rombach.pdf

A. A. Romeo (1975), 'Interindustry and interfirm differences in the rate of diffusion of an innovation', *Review of Economics and Statistics* 57, 311–19.

M. Rönn, J. E. Andersson, and G. B. Zettersten (eds.) (2013), *Architectural Competitions: Histories and Practice*. Stockholm: Royal Institute of Technology and Rio Kulturkooperativ.

S. Roper, C. Hales, J. R. Bryson, and J. Love (2009), *Measuring Sectoral Innovation Capability in Nine Areas of the UK economy*. London: Nesta.

S. Roper and H. Hewitt-Dundassa (2015), 'Knowledge stocks, knowledge flows and innovation: Evidence from matched patents and innovation panel data', *Research Policy* 44 (7), 1327–40.

S. Roper and J. H. Love (2002), 'Innovation and export performance: evidence from the UK and German manufacturing plants', *Research Policy* 31 (7), 1087–102.

S. Roper, P. Micheli, J. H. Love, and P. Vahter (2015), 'The roles and effectiveness of design in new product development: A study of Irish manufacturers', *Research Policy* 45 (1), 319–29.

P. R. Rosenbaum and D. B. Rubin (1985), 'Constructing a control group using multivariate matched sampling methods that incorporate the propensity score', *American Statistician*, 39 (1), 33–8.

N. Rosenberg (1976), 'On technological expectations', *Economic Journal*, 86 (343), 523–35.

P. Rouvinen (2002), 'R&D-productivity dynamics: Causality, lags, and dry holes', *Journal of Applied Economics* 5 (1), 123–56.

R. P. Rumelt (1991), 'How much does industry matter?' *Strategic Management Journal* 12 (3), 167–85.

S. Salop (1979), 'Monopolistic competition with outside goods', *Bell Journal of Economics* 10, 141–56.

F. M. Scherer (1982), 'Inter-Industry technology flows and productivity growth', *Review of Economics and Statistics* 64, 627–34.

R. Schmalensee (1985), 'Do markets differ much?' *American Economic Review* 75 (3), 341–51.

T. Schmidt and C. Rammer (2007), *Non-Technological and Technological Innovation: Strange Bedfellows?* (Discussion Paper 07-052). ZEW-Centre for European Economic Research. http://ftp.zew.de/pub/zew-docs/dp/dp07052.pdf

S. P. Schnaars (1994), *Managing Imitation Strategies: How Late Entrants Seize Marketing from Pioneers*. New York: Free Press.

T. Schubert (2010), 'Marketing and organisational innovations in entrepreneurial innovation processes and their relation to market structure and firm characteristics', *Review of Industrial Organization* 36 (2), 189–212.

J. A. Schumpeter (1934), *The Theory of Economic Development: An Inquiry into Profits, Capital, Credit, Interest, and the Business Cycle*. New Brunswick, NJ: Transaction.

K. Seim (2006), 'An empirical model of firm entry with endogenous product type choices', *RAND Journal of Economics* 37 (3), 619–40.

A. Shaked and J. Sutton (1982), 'Relaxing price competition through product differentiation', *Review of Economic Studies* 49 (1), 3–13.

A. Shaked and J. Sutton (1983), 'Natural monopolies', *Econometrica* 51 (5), 1469–84.

A. Shaked and J. Sutton (1987), 'Product differentiation and industrial structure', *Journal of Industrial Economics* 36 (2), 131–46.

D. Skuras, K. Tsegenidi, and K. Tsegouras (2008), 'Product innovation and the decision to invest in fixed capital assets: Evidence from an SME survey in six European Union member states', *Research Policy* 37 (10), 1778–89.

S. F. Slater and J. C. Narver (1994), 'Does competitive environment moderate the market orientation-performance relationship?' *Journal of Marketing* 58, 46–55.

M. Smirlock, T. Gilligan, and W. Marshall (1984), 'Tobin's Q and the structure-performance relationship', *American Economic Review* 74 (5), 1051–60.

A. T. Sorensen (2007), 'Best seller lists and product variety', *Journal of Industrial Economics* 55 (4), 715–38.

P. R. Steffens (2003), 'A model of multiple ownership as a diffusion process', *Technological Forecasting and Social Change* 70 (9), 901–17.

A. Sterlacchini (1999), 'Do innovative activities matter to small firms in non-R&D-intensive industries? An application to export performance', *Research Policy* 28 (8), 819–32.

J. Stiglitz and A. Weiss (1981), 'Credit rationing in markets with imperfect information', *American Economic Review* 71, 393–410.

P. Stoneman (1981), 'Intra-firm diffusion, Bayesian learning and profitability', *Economic Journal* 91, 375–88.

P. Stoneman (1983), *The Economic Analysis of Technological Change*. Oxford: Oxford University Press.

P. Stoneman (1987), *The Economic Analysis of Technology Policy*. Oxford: Oxford University Press.

P. Stoneman (1989), 'Technological diffusion, vertical product differentiation and quality improvement', *Economic Letters* 31, 277–80.

P. Stoneman (1990), 'Technological diffusion, horizontal product differentiation and adaptation costs', *Economica* 57, 49–62.

P. Stoneman (1991), 'The use of a levy/grant scheme as an alternative to tax based incentives to R & D', *Research Policy* 20, 195–201.

P. Stoneman (ed.) (1995), *Handbook of the Economics of Innovation and Technological Change*. Oxford: Basil Blackwell.

P. Stoneman (2002), *The Economics of Technological Diffusion*. Oxford: Basil Blackwell.

P. Stoneman (2011), *Soft Innovation: Economics, Product Aesthetics, and the Creative Industries* (rev. edn). Oxford: Oxford University Press.

P. Stoneman (2013), 'The impact of prior use on the further diffusion of new process technology', *Economics of Innovation and New Technology* 22 (3), 238–55. doi: 10.1080/10438599.2012.708133.

P. Stoneman and G. Battisti (2010), 'The diffusion of new technology', in B. Hall and N. Rosenberg (eds.), *Handbook of the Economics of Innovation*, vol. 2 Amsterdam: Elsevier, 734–60.

P. Stoneman and M. Karshenas (1992), 'A flexible Model of technological diffusion incorporating economic factors with an application to the spread of colour television ownership in the UK', *Journal of Forecasting* 11 (7), 577–601.

P. Stoneman and M. J. Kwon (1994), 'The diffusion of multiple process technologies', *Economic Journal* 104, 420–31.

P. Stoneman, V. Wong, and W. Turner (1996), 'Marketing strategies and market prospects for environmentally friendly consumer products', *British Journal of Management* 7 (3), 263–81.

A. K. Sundaram, T. A. John, and K. John (1996), 'An empirical analysis of strategic competition and firm values. The case of R&D competition', *Journal of Financial Economics*, 40 (3), 459–86. doi: 10.1016/0304-405X(95)00853-7.

J. Sutton (1986), 'Vertical product differentiation: some basic themes', *American Economic Review*, Papers and Proceedings, 76, 393–8.

J. Sutton (1998), *Technology and Market Structure, Theory and History*, London: MIT Press.

T. Takalo, T. Tanayama, and O. Toivanen (2013), 'Estimating the benefits of targeted R&D subsidies', *Review of Economics and Statistics* 95, 255–27.

D. J. Teece, G. Tsano, and A. Shuen (1997), 'Dynamic capabilities and strategic management, *Strategic Management Journal* 18(7), 509–33.

B. S. Tether (2006), *Design in Innovation: Coming out from the Shadow of R&D, An analysis of the UK Innovation Surveys of 2005*, London: Department for Innovation, Universities and Skills, HM Government.

J. Tidd (ed.) (2012), *From Knowledge Management to Strategic Competence: Assessing Technological Market and Organisational Innovation* (3rd edn) (Technology and Management 19). London: Imperial College Press.

J. Tirole (1988), *The Theory of Industrial Organization*. Cambridge, MA: MIT Press.

O. Toivanen, P. Stoneman, and D. Bosworth (2002), 'Innovation and the market value of UK firms, 1989–1995', *Oxford Bulletin of Economics and Statistics* 64 (1), 39–61.

D. M. Topkis (1978), 'Minimizing a submodular function on a lattice', *Operations Research* 26 (2), 305–21.

M. Trajtenberg (1989), 'The welfare analysis of product innovations with an application to computed topography scanners', *Journal of Political Economy* 97 (2), 444–79.

J. E. Triplett (1982), 'Concepts of quality in input and output price measures: A resolution of the user-value resource-cost debate', in M. F. Foss (ed.), *The US National Income and Product Accounts: Selected Topics*. Chicago, IL: University of Chicago Press, 269–312.

J. E. Triplett (2006), 'The Boskin Commission report after a decade', *International Productivity Monitor* 12, 42–60.

The United Nations Conference on Trade and Development [UNCTAD] (2017) , *Global Invetsments Trend Monitor*, 1 February. http://unctad.org/en/PublicationsLibrary/webdiaeia2017d1_en.pdf

UNESCO Institute for Statistics [UIS] (2012a), *Results of the 2011 UIS Pilot Data Collection of Innovation Statistics*. Quebec: UNESCO Institute for Statistics.

UNESCO Institute for Statistics [UIS] (2012b), 'From international blockbusters to national hits', *UIS Information Bulletin* 8, 1–24. http://unesdoc.unesco.org/images/0021/002171/217103e.pdf

UNESCO Institute for Statistics [UIS] (2013), 'Feature film diversity', UIS Fact Sheet, No. 24, May. http://unesdoc.unesco.org/images/0022/002216/221628e.pdf

J. M. Utterback (1994), *Mastering the Dynamics of Innovation*. Boston, MA: Harvard Business School Press.

J. M. Utterback and W. J. Abernathy (1975), 'A dynamic model of process and product innovation', *OMEGA* 3 (6), 639–56.

E. Uyarra, J. Edler, S. Gee, L. Georghiou, and J. Yeow (2014), 'Public procurement of innovation: The UK case', in V. Lember et al. (eds.), *Public Procurement, Innovation and Policy*. Berlin: Springer. doi: 10.1007/978-3-642-40258-6_12.

B. Van Ark, R. Inklaar, and R. H. McGuckin (2003), 'ICT and productivity in Europe and the United States: Where do the differences come from?', *CESifo Economic Studies* 49 (3), 295–318.

K. Van Drongelen, C. Inge, and A. Cooke (1997), 'Design principles for the development of measurement systems for Research and Development processes', *R&D Management* 27 (4): 345–57.

P. A. VanderWerf and J. F. Mahon (1997), 'Meta-analysis of the impact of research methods on findings of first-mover advantage', *Management Science* 43 (11), 1510–9.

T. Veblen (1899), *The Theory of the Leisure Class*. New York: Macmillan.

N. Venkatraman (1989), 'Strategic orientation of business enterprises: The construct, dimensionality and measurement', *Management Science* 35 (8), 942–62.

R. Vernon (1966), 'International investment and international trade in the product cycle', *Quarterly Journal of Economics*, 80, 190–207.

K. Wakelin (1998), 'Innovation and export behaviour at the firm level', *Research Policy* 26 (7–8), 829–41.

J. Wells and A. Restieaux (2014), *Review of Hedonic Quality Adjustment in UK Consumer Price Statistics and Internationally*. London: Office for National Statistics.

B. Wernerfelt (1984), 'A resource-based view of the firm', *Strategic Management Journal* 5 (2), 171–80.

R. Whittington, A. Pettigrew, S. Peck, E. Fenton, and M. Conyon (1999), 'Change and complementarities in the new competitive landscape: A European panel study, 1992–1996', *Organization Science* 10 (5), 583–600.

R. Winger and G. Wall (2006), *Food Product Innovation: A Background Paper*. Rome: Food and Agricultural Organisation of the United Nations.

World Intellectual Property Organisation [WIPO] (2009), *WIPO Gazette of International Marks: Statistical Supplement for 2009*. Geneva: International Bureau of the World Intellectual Property Organization.

World Intellectual Property Organization [WIPO] (2011), *What Is Intellectual Property?*, Geneva: WIPO Publication No. 450 (E). http://www.wipo.int/edocs/pubdocs/en/intproperty/450/wipo_pub_450.pdf

World Intellectual Property Organisation [WIPO] (2012), 'The rise of design in innovation and intellectual property: Definitional and measurement issues', in *World Intellectual Property Indicators 2012 Edition*. Economics and Statistics Division, Geneva. www.wipo.int/ipstats

World Intellectual Property Organisation [WIPO] (2013), *2013 Madrid Yearly Review*. Geneva: WIPO Economics and Statistics Series.

INDEX

Tables and figures are indicated by an italic *t* and *f* following the page number.

Abernathy, W. J. 57, 58
Acemoglu, D. 1
ACNielsen 55, 169
Adams, R. 137
adverse selection
  financial constraints to innovation 106
  insurance coverage, absence of 102
advertising
  demand 64, 69
  new-to-market innovations 86
  original products 76
Afonso, Ó. 132
Aghion, P. 16, 202
Ahnert, H. 172
Airbus 212
Alange, S. 129
Allen, F. 102
Anderson, E. W. 162
Añón Higón, D. 185
Antonelli, C. 129
Antoniades, A. 153
Apple 78–9
architectural competitions 219
Arrow, K. 93, 101, 190
artistic design 19
Arts Council 208
Association of the British Pharmaceutical Industry 211
Atuahene-Gima, K. 164, 165
Audretsch, D. B. 114
Australia
  extent of product innovation 47, 53
  food industry 47
  music 53
Austria, extent of product innovation 33, 34*t*, 35*t*, 37*t*

Bain, A. D. 132
Baines, P. 10
Baker, W. E. 165
Ballot, G. 157
Baptista, R. 124
Barak, L. 129
barriers to innovation 88–9, 103–7, 117–19, 118*t*
Bartoloni, E. 110, 158
Bass, F. M. 124
Battisti, G. 16, 33, 63, 109, 110, 111, 115, 116, 124, 129, 131–2, 152, 158, 159
Baussola, M. 110, 158
Becker, B. 113, 114
Becker, S. O. 154
Belgium
  extent of product innovation 33, 34*t*, 35*t*, 37*t*
  patent box 205
Beneito, P. 196
Benston, G. J. 139
Bernard, A. B. 153
Berndt, E. R. 178
Besanko, D. 102
Binmore, K. 213
Blanchard, O. J. 169, 170
Blind, K. 215
Blonigen, B. A. 121
Bloom, N. 126, 203 n. 2
Boeing 212
Bohlin, A. 125
Bond, S. 106, 107
book publishing 51–3, 52*t*
Borin, A. 153
Boskin, M. J. 171–2, 177, 178
Bradsher, K. 78
branding 10–11
  extent of product innovation 111, 112
  firm performance 148–9, 163
Brauetigam, R. R. 102
Brazil
  extent of product innovation 43, 44*t*, 53
  music 53
Brealey, R. 102
Bresnahan, T. F. 12, 14
Broda, C. 54–5, 56, 60, 169–70, 172, 174, 178
Brouwer, E. 115
Brown, M. G. 137
Brown, S. L. 137
Bulgaria, extent of product innovation 34*t*, 35*t*, 37*t*
business processes 111, 112, 113

Bustos, P. 153
Buzzell, R. D. 161

Cadotte, E. R. 162
Caldera, A. 153
Camus, D. 175
Canada
  extent of product innovation 53
  music 53
Canepa, A. 105
capacity creation 73, 87
  interdependencies 73, 77–8
  new-to-market innovations 80, 84, 86
  original products 76–8
  outsourcing 79
capacity location *see* location of production capacity
capital intensity 121
Cassiman, A. 154–5, 194
Cassiman, B. 22
Castellion, G. 13, 44, 101
Catapult centres 213
Cefis, E. 137
Chandrasekaran, D. 124
Chaney, P. K. 148
Chang, T. Z. 142
characteristics approach to consumer behaviour 175
Chen, S.-S. 149
Chen, Su-Jane 142
Chesbrough, H. W. 22
China
  Apple 78
  book publishing 52, 52*t*
  diffusion of production of product innovations 133, 134
  extent of product innovation
    product launch data 48, 50, 50*t*, 52, 52*t*
    surveys 43, 44*t*
  film industry 50, 50*t*
  food industry 48
  foreign direct investment 215–16
  nuclear energy 210
  outsourcing 80
Ciccarelli, M. 137
Cincera, M. 120
Clark, G. 124, 129
Cohen, W. 113, 114
Colombia, extent of product innovation 43, 44*t*
Comin, D. A. 77, 121–2, 126–8, 129, 130
commodity markets 10
common pool effects 85
  extent of product innovation 94, 98
  intellectual property rights 202
  prizes 219
  welfare 189
community design rights 27, 198
Community Innovation Surveys (CIS) 13 n. 3
  constraints on innovation 103, 105–6, 118
    financial 203
  definition of product innovation 13
  design activities 20
  extent of product innovation 33–8, 34*t*, 35*t*, 36*t*, 37*t*, 39, 111
    comparison with US innovation survey 43
  firm performance
    estimates of impact of product innovation on 150
    innovation complementarities 157, 158
    marketing innovation 157
    organizational innovation 157
  limitations of data 44, 46
  tax incentives 203
  UK *see* United Kingdom: Innovation Surveys
competition
  as constraint on innovation 118
  as determinant of innovativeness 113, 114
  firm performance 136, 159–60
  policy dimension 195–6, 202
  welfare 183–4, 185, 192–3
complementarities, innovation 156–9
computer ownership subsidies 217
confidentiality agreements 200
Connolly, R. A. 139
constraints on innovation 88–9, 103–7, 117–19, 118*t*
consumer price index (CPI) 55, 170, 174
  Boskin Commission 171–2
  EU 172
  hedonic methods 175–8
consumer products 125–6
  defined 9
  durables 67–9, 127
  non-durables 68–9
consumer surplus 203 n. 2
  intellectual property rights 200, 201, 203
  welfare 180–91, 181*f*
Cook, A. 142
Cooper, R. 160
copyright 26, 28–9
  policy dimension 197, 198, 200, 201
corporation tax 203–4, 205, 216
Corsino, M. 115

cost constraints *see* financial constraints to innovation
cost-of-living index (COLI) 170, 172
  hedonic methods 177–8
Cozzarin, B. P. 16, 110
Crass, D. 163
creative destruction 202
creativity 19, 20–1
  defined 20
  incentives to product innovation 90, 91
  subsidies 208
  see also book publishing; film industry; music
credit constraints *see* financial constraints to innovation
Crepon, B. 145, 150–1, 166
Croatia, extent of product innovation 34*t*, 35*t*, 37*t*
Cross Country Historical Adoption of Technology (CHAT) data set 127, 129
Crozet, M. 156, 161
Cusumano, M. A. 70
Cyprus, extent of product innovation 34*t*, 35*t*, 37*t*
Czech Republic, extent of product innovation 34*t*, 35*t*, 37*t*

Dacko, S. G. 25
Damijan, J. P. 155
Dasgupta, P. 16, 190, 202
d'Aspremont, A. J. 162
David, P. A. 16, 208, 215
Davila, T. 137
Davis, C. R. 161
Davis, S. J. 169, 170
Day, G. S. 163
Deaton, A. 178 n. 4
Dechezleprêtre, A. 204
Defence Committee of the House of Commons 210
defence sector 208–10
defensive innovation 89, 90
de Jong, J. P. J. 113
demand for new products 62–72
  consumer durables and non-durables 67–9
  new-to-firm innovations 71
  new-to-market innovations 69–71
  original, producer durable, monopoly supplier 63–7
demand pull/push factors 89, 90, 91
democracy 195
Denmark, extent of product innovation 33, 34*t*, 37*t*
Department for Business, Energy and Industrial Strategy (DBEIS) 19, 38, 77
Department for Business Innovation and Skills (DBIS) 22, 88, 199
Department of Culture, Media and Sport (DCMS) 20
deregulation *see* regulations
Deshpande, R. 163
design 16, 19–20
  extent of product innovation 111, 112–13
design rights (design patents) 26, 27–8, 28*t*
  policy dimension 197–8, 200
  R&D spending 30*t*, 30
determinants of innovativeness 112–17
development spending 16, 17, 18, 91
Dickerson, M. D. 129
Diewert, W. E. 173
differentiated products *see* product differentiation
diffusion of product innovations 124–35
  extensive margin 126–8
  intensive margin 128–30
    intra-firm and intra-household diffusion 131–2
    production of product innovations 132–4
Dimos, C. 207
Dixit, A. K. 65, 84, 97, 103, 192
Doran, J. 157
Dosi, G. 12
Drazin, R. 105
Duguet, E. 145, 150–1, 166
Duhigg, C. 78
du Plessis, M. 24
durable products
  consumer durables 67–9, 127
  diffusion of 126, 131
  extent of product innovation 98–9
  incentives to product innovation 89–90
  intellectual property rights 202
  pricing 74–5, 81–2, 85
  producer durables *see* producer products: durables
  welfare 185–9, 190
dynamic capabilities approach 164

early-mover advantages/disadvantages 25
  demand for new products
    consumer durables 67
    producer durables 65–6
East Asian tigers 127
Eaton, J. 132

economic profit to capital employed (EP/CE) ratio 145
economic value added (EVA) 139
Economides, N. 162
Edgett, S. 160
Edquist, C. 194
Edwards, J. 139
Egger, P. H. 154
Egypt, extent of product innovation 43, 44*t*
Eisenhardt, K. M. 137
electric vehicles
  charging stations 214
  subsidies 217
emulation 16, 24–30
  product differentiation 162
  welfare 185
engineering design 19
Ennen, C. 156
Enterprise Investment Scheme (EIS) 211–12
entry to industries 58–9, 119–21
Epstein, M. 137
equipment, sourcing new 111, 112
Estonia, extent of product innovation 34*t*, 35*t*, 37*t*
Ethiraj, S. K. 162
European Commission 13 n. 3
European Patent Office (EPO) 27
  patent box 205
European Union (EU)
  Airbus 212
  Community Innovation Surveys 13 n. 3
  Community Trade Mark 29, 199
  constraints on innovation 118–19, 119*t*
    risk and finance 105, 107
  copyright 29, 198, 201
  design rights 27, 28*t*, 198
  diffusion of production of product innovations 133
  exhaust emission regulations 214
  extent of product innovation 43
  Intellectual Property Office 29, 199
  Orphan Works Directive 201
  patents 27, 197
  price indexes 172–3, 179
    hedonic methods 176, 177
  surveys of product innovation 32–41, 34*t*, 35*t*, 36*t*, 37*t*
  trademarks 199
Eurostat
  price indexes 168, 170, 172, 177
  surveys of product innovation 32–3
exhaust emission regulations 214
exit from industries 58–9, 120, 121
exports
  as determinant of innovativeness 114, 117
  diffusion of product innovations 126
  firm performance 141, 152–6
  intellectual property rights 203
  of original products 78
  welfare 184–5
extensive margin of technology diffusion 126–8, 129
extent of product innovation 32–61, 91–4
  empirical evidence on the determination of 108–23
    constraints on innovation 117–19
    determinants of innovativeness 112–17
    entry, start-ups, foreign direct investment, and imports 119–22
    survey data 109–12
  horizontal product innovation 94–9
  industry maturity 57–9
  product launch data 44–54
  scanner data 54–6
  surveys 32–44
  vertical product innovation 99–101

factor scores 112
Fama, C. F. 148, 148 n. 4
Fantino, D. 192
Faraday Partnerships 213
Feder, G. 129
Federation of European Publishers 52
Feenstra, R. C. 79
Fehle, C. 148
Feigl, S. 47
Feldman, M. P. 124
Ferreira, R. D.-S. 162
Filipetti, A. 20
film industry
  product launch data 48–51, 50*t*
  tax incentives 206
final products 9
finance as determinant of innovativeness 114
financial constraints to innovation 104, 105–7, 117–18, 118*t*, 119
  policy dimension 203, 211–12
  tax incentives 203, 207
Finland
  extent of product innovation 33, 34*t*, 35*t*, 37*t*, 38
  launch aid 212
  subsidies 208
firm characteristics, as determinant of innovativeness 114, 116–17

firm-level product innovation *see* new-to-firm innovations
firm performance 136–67
estimates of impact of product innovation on 142–56
innovation complementarities 156–9
market orientation and product development process 163–5
measurement 138–42
accounting-based measures 138–9, 140
cost accounting measures 140–1
market-based measures 140
non-financial measures 142
product differentiation 161–3
product portfolios and life cycles 159–61
firm productivity, measurement 141
firm size
as determinant of innovativeness 113–14, 115, 116
extent of product innovation 93
UK 38, 38*t*, 39, 40*t*
financial constraints to innovation 106
patents 200
and performance 148
first-mover effects 25
demand for new products
consumer durables 67
producer durables 65–6
fiscal policy 195
public sector borrowing requirement 206
Fisher, F. M. 139, 146
Flaherty, M. T. 161
follower firms 25
see also emulation
food industry 47–8, 48*t*
foreign direct investment (FDI) 16, 24
extent of product innovation 119, 121
original products 78
policy dimension 215–17
France
diffusion of production of product innovations 133
extent of product innovation
product launch data 53
surveys 33, 34*t*, 35*t*, 36*t*, 36, 37*t*
firm performance
estimates of impact of product innovation on 150
innovation complementarities 158
market value 147
music 53
nuclear energy 210
patent box 205
Frantzen, D. 152
Fraunhofer Society 213
Freeman, C. 89
French, K. R. 148, 148 n. 4
Frenz, M. 110
Fritsch, M. 115
Fudenberg, D. 65

Gabszewicz, J. 10, 69, 90
Galgau, O. 120
Ganotakis, P. 154
Gao, Z. 132
Gatignon, H. 164, 165
Gentry, J. W. 129
Germany
Apple 78
diffusion of production of product innovations 133, 134
extent of product innovation
product launch data 47, 48, 53
surveys 33, 34*t*, 36*t*, 36, 37*t*, 38
firm performance
estimates of impact of product innovation on 151, 153–4
innovation complementarities 157
market value 147
product differentiation 163
food industry 47, 48
Fraunhofer Society 213
music 53
patent box 205
price indexes 177
subsidies 218
Geroski, P. 26, 57, 136, 143–4
Gilligan, T. 139
Glaister, S. 69, 75, 76
global innovation, defined 11, 12
global perspective, diffusion of product innovations 126–8
goods
extent of product innovation 33–5, 35*t*
and services, distinction between 8
Gordon, R. J. 12
Gort, M. 120
government acquisition 208–11
Grady, E. 58, 59
grant-based policies 207–8
Greece, extent of product innovation 33, 34*t*
Greenley, G. E. 142
Greenstein, S. 101, 192, 215
Griliches, Z. 32, 129
Grossman, G. M. 152, 202
growth theory 1

Hall, B. 106, 146, 147, 150, 151, 166, 190, 199, 200, 203 n. 2
Hall, W. K. 161
Haltiwanger, J. 169, 170
Han, J. K. 164–5
Hanson, G. H. 79
Harhoff, D. 199
Harmonised Index of Consumer Prices (HICP) 172
Harris, L. C. 142
Harrison, John 218–19
Hauser, J. 137
Hausman, J. A. 176–8, 184
Hawawini, G. 144
Hawking, S. vi
health and safety regulations 215
hedonic regression 174, 175–8
Heffernan, S. A. 41
Helpman, E. 152, 202
Hendricks, K. B. 149
Herguera, I. 207
Heshmati, A. 145
Hewitt-Dundas, H. 22
Historical Cross Country Technology Adoption (HCCTAD) data set 127, 128, 129
Hobijn, B. 126–8, 129, 130
horizontal product differentiation 10
  branding 11
  defined 14
  demand for new products 69, 70
  extent of 94–9
  firm performance 161–3
  new-to-market innovations 80–2, 84–5
  welfare 182–3, 192
Hotelling, H. 10, 69, 94, 162
Howitt, P. 202
Hubbard, R. G. 106
Hult, G. T. M. 164
human capital *see* labour force
Hungary, extent of product innovation 34*t*, 35*t*, 37*t*, 38
Hurley, R. F. 164

Iceland, extent of product innovation 34, 35*t*, 36, 37*t*
Ichniowski, C. 159
Imbriani, C. 155
imitation *see* emulation
implicit quality indexes (IQIs) 174
imports 16, 24
  diffusion of product innovations 126
  extent of product innovation 119, 121–2
  intellectual property rights 203
  subsidies 218
  welfare 184–5
incentives to product innovation 88–91, 107
incremental innovation 159
India
  book publishing 52*t*
  extent of product innovation 50, 50*t*, 52*t*
  film industry 50, 50*t*
Indonesia
  book publishing 52*t*
  extent of product innovation 52*t*
industrial activities, classification of 8–9
industrial design 19
industries
  characteristics
    as determinant of innovativeness 116
    extent of product innovation 121
  defined 8
  effect of product innovation upon 14–15
  entry and exit 58–9
  life cycle 57–9
    as determinant of innovativeness 114
    firm performance 160
    new firm entry 120
  maturity 57–9
infrastructure 213–14
Inklaar, R. 177 n. 3
Innovate UK 213
innovation effectiveness curve 141
Institute for Economic Research (IFO) 154
insurance coverage/markets 102, 105
Intellectual Property Office (EU) 29, 199
Intellectual Property Office (UK) 28, 198–9, 201, 205
  Trade Marks Register 29, 198–9
intellectual property rights (IPR) 25–30
  extent of product innovation 32, 47
  outsourcing 80
  policy dimension 196–203
  public contracts 210
  R&D spending 29–30, 30*t*
intensive margin of technology diffusion 127, 128–34
  intra-firm and intra-household diffusion 131–2
  production of product innovations 132–4
Interbrand 148
inter-firm diffusion of product innovations 129, 131–2
intermediate products 9
International Federation of the Phonographic Industry (IFPI) 53–4

intra-firm diffusion of product innovations 64, 129, 131–2
  early-mover advantages 65
intra-household diffusion of product innovations 129, 131–2
Iona, A. 132, 158
Ireland
  extent of product innovation 33, 34*t*, 35*t*, 36*t*, 37*t*
  foreign direct investment 216–17
  patent box 205
Ireland, N. 75, 188–9
Irish Development Agency (IDA) 216
Israel, extent of product innovation 43, 44*t*
Italy
  diffusion of production of product innovations 133
  extent of product innovation 33, 34*t*, 35*t*, 36, 37*t*, 38
  firm performance
    estimates of impact of product innovation on 155–6
    innovation complementarities 158
    market value 147
  tax incentives 206

Jaffe, A. 205
Jang, S. S. 162
Japan
  Apple 78
  diffusion of product innovations 127, 133, 134
  extent of product innovation 50*t*, 53, 54
  film industry 50*t*
  music 53, 54
Jaumandreu, J. 151, 184, 184 n. 2
Jenkins, R. L. 162
Jensen, J. Bradford 153
Jensen, R. 64
Jewkes, J. 18
Jing, B. 162
Johnson, D. S. 176
joint ventures 80
Jones, C. 202

Kaiser, U. 76
Kaivanto, K. 212
Kandybin, A. 140, 141
Karshenas, M. 72, 132
Keely, L. 159 n. 1
Kenny, G. 172
Kerssens-van Drongelen, I. C. 142
Kleinknecht, A. 115
Kleinschmidt, E. 160
Klemperer, P. 213
Klepper, S. 58, 59, 114, 120, 159
knowledge
  constraints on innovation 104–5, 117, 118*t*, 118
  flows 16, 21–4
    sources 22–4, 23*t*
  policies to facilitate access to 213
Kogler, D. F. 124
Köhler, C. 204
Korea
  Apple 78
  diffusion of production of product innovations 133, 134
  extent of product innovation 48
  firm performance 162–3
  food industry 48
Kortum, S. 132
Krishnan, V. 137
Kuczmarski, T. D. 140
Kwon, M. J. 129

labour force
  as constraint on innovation 104–5, 118, 119
  extent of product innovation 113, 114, 116, 117
  location of production capacity 79
Lambert, R. 110
Lambertini, L. 193
Lambkin, M. 25, 65
Lamperti, F. 22
Lancaster, K. J. 175, 192
Lanzolla, G. 25, 66, 103
Laspeyres price index 171
Latvia, extent of product innovation 33, 34*t*, 34, 35*t*, 37*t*
launch aid 212
laundry detergents, product launch data 46
law, rule of 195
Lerner, J. 106
Levin, R. 200
levy grant system 206
Lhuillery, S. 158
Libai, B. 131
Lieberman, M. B. 25, 65
Liikanen, J. 125, 129
Lin, E. S. 216
Lin, H. L. 216
Lin, P. 93
Lindenberg, E. B. 139
Lindsay, D. 201

Lithuania, extent of product innovation 34*t*, 35*t*, 37*t*
local product innovation *see* new-to-market innovations
location of production capacity 73, 87
  interdependencies 73, 77–8
  labour factors 104–5
  original products 76–8
Lofsten, H. 160
Longitude Act (1714) 218–19
Loof, H. 145
Love, J. H. 153, 154
Lundvall, B.-Å. 1
Lutz, S. 101, 207
Luxembourg, extent of product innovation 33, 34*t*, 34, 35*t*, 37*t*, 38

machinery innovation
  complementarities 158
  extent of 109, 110
macroeconomics of innovation 1
Madakom 47
Madrid trademark system 47, 199
  food industry 47, 48*t*, 48
Mahajan, V. 124
Mahon, J. F. 25, 65
Mairesse, J. 145, 150–1, 166, 184, 184 n. 2
Malaysia
  diffusion of production of product innovations 133
  extent of product innovation 43, 44*t*
Malta, extent of product innovation 34*t*, 35*t*, 37*t*
management innovation
  complementarities 158
  demand for a new product 63
  diffusion 126
  extent of 109, 110
  surveys 33
Mansfield, E. 63, 64, 131
marketing
  firm performance 164
  innovation
    complementarities 157–8
    defined 157
    extent of 108, 109, 110, 111, 112
    firm performance 165
    surveys 33
  product differentiation 163
market orientation 163–5
markets
  characteristics
    constraints on innovation 117, 118*t*, 119
    as determinant of innovativeness 114
    extent of product innovation 120–1
  classification 9
  constraints to innovation 104
  defined 8, 9
  failure 195
  incentives to product innovation 89, 90, 91
  size 93
  structure 93
market-to-book value of the firm 140
market value added 140
market value of firms 146–9
Markham, S. K. 13, 44, 101
Markowitz, H. M. 65
Marshall, W. 139
McCall, B. 24
McGowan, J. J. 139, 146
McGuckin, R. H. 177 n. 3
McNamara, K. T. 47
Melitz, M. J. 153, 155
Menrad, K. 47
Meschede, M. 115
Mestieri, M. 77, 121–2, 127, 129
Mexico, diffusion of production of product innovations 133
Milgrom, P. 156, 157
Milling, P. 58
Mina, A. 106
Mlodinow, L. vi
mobile telephony, infrastructure 213
Mohnen, P. 157
momentum index 148
monetary policy 195
monopolies
  demand for new products
    consumer durables and non-durables 67–9
    producer durables 63–7
  extent of product innovation 113, 114
    horizontal product innovation 94, 97, 98, 99
    vertical product innovation 99–101
  intellectual property rights 200–1, 202
  new-to-firm innovations 84
  new-to-market innovations
    horizontal product differentiation 80–2
    promotion and capacity creation 83–4
    vertical product differentiation 82–3
  original products
    capacity creation 76–8
    capacity location 78–80

pricing 74–5
promotion 76
welfare 182–3, 189–90, 192–3
Montgomery, D. B. 25, 65
Moon, Y. 160
moral hazard
financial constraints to innovation 106
insurance coverage, absence of 102
Moran, T. H. 121
Moretti, E. 209
Morris, R. 19
Moschini, G. 101
Mostacci, F. 173 n. 2
Mueller, D. C. 120
Mumford, M. D. 20
music 53–4
Myers, S. 102

Nagaoka, S. 32
Narver, J. C. 142, 163, 164, 165
National Audit Office 210
National Endowment for the Science Technology and the Arts (NESTA) 111
National Health Service (NHS) 211
National Institute of Health and Care Excellence (NICE) 211
National Science Foundation, Business R&D and Innovation Survey (BRDIS) 41–3, 42*t*
Nelson, R. R. 104, 118
Netherlands
electric car charging stations 214
extent of product innovation 33, 34*t*, 35*t*, 36*t*, 36, 37*t*
launch aid 212
network effects, demand for new products
consumer durables 67, 68
new-to-market innovations 71
producer durables 65
Neven, D. 162
new products
defined 12
survival rate 141
see also original products
new-to-firm innovations
defined 11–12
demand for 71
diffusion 124–5
emulation 24–30
extent of 110
intertemporal pattern 12
monopolies 84
oligopolies 86–7
surveys 36–8, 37*t*
UK 39, 40–1*t*
new-to-market innovations
defined 11
demand for 69–71
diffusion 124–5
extent of 94–7, 110
intertemporal pattern 12
monopolies
horizontal product innovation 80–2
promotion and capacity creation 83–4
vertical product innovation 82–3
oligopolies
horizontal product differentiation 84–5
promotion and capacity creation 86
vertical product differentiation 85–6
surveys 36–8, 37*t*
UK 39, 40–1*t*
New Zealand
extent of product innovation 47
food industry 47
Nigeria
extent of product innovation 50, 50*t*
film industry 50, 50*t*
Noh, Y.-H. 101
non-disclosure agreements 200
non-durable products 126
consumer non-durables 68 9
diffusion of 131
pricing 74, 75, 80–1, 84–5
welfare 180–5, 189–90
North–South models, diffusion of production of product innovations 132
Norway, extent of product innovation 33, 34*t*, 35*t*, 37*t*

offensive innovation 89
oligopolies
extent of product innovation
horizontal product innovation 94–7, 98, 99
vertical product innovation 100, 101
new-to-firm innovations 86–7
new-to-market innovations
horizontal product differentiation 84–5
promotion and capacity creation 86
vertical product differentiation 85–6
Olson, E. M. 164
O'Mara, G. T. 129
open innovation 22
option costs 174–5

Organisation for Economic Co-Operation and Development (OECD)
  definition of product innovation 13, 14
  design activities 19
  diffusion of product innovations 127, 128
  extent of product innovation 44
  firm performance 152
  knowledge flows 22
  non-technological innovation 157
  price indexes 179
  R&D 17, 18–19, 206
  tax incentives 203
organizational innovation
  complementarities 157, 158
  defined 157
  demand for a new product 63
  determinants of 114
  extent of 108, 109, 110–11
  surveys 33
Oriani, R. 147
original products
  capacity creation 76–8
  capacity location 78–80
  defined 12
  demand for 63–7
  diffusion 124–5
  intertemporal pattern 12
  price indexes 174
  pricing 74–5
  promotion 76
orphan works 201
Orsini, R. 193
Ortmans, L. 20
outsourcing 78–80
overseas direct investment (ODI) *see* foreign direct investment (FDI)

Paasche price index 171
PA Consulting Ltd 102
Pakistan
  book publishing 52*t*
  extent of product innovation 52*t*
Park, K. M. 162
patent box 205–6
patents 26–7, 28, 29
  design *see* design rights
  estimates of impact of product innovation on 146, 147
  extent of product innovation 32, 47
  knowledge flows 21, 22
  policy dimension 197, 198, 199, 200
  R&D spending 29–30, 30*t*
  tax credits 204, 205
  utility 26
  welfare 190
Penrose, E. T. 164
Peres, R. 124, 129
performance *see* firm performance
Peri, G. 21
Peters, B. 151
Petrin, A. 151
Pharmaceutical Price Regulation Scheme (PPRS) 211
pharmaceuticals
  product launch data 44–6, 45*t*
  R&D 17–18
Philippines, extent of product innovation 43, 44*t*
Phillips, L. W. 162
Piger, J. 121
Pindyck, R. 65, 103
Pisano, G. P. 160
Pitkethly, R. 200
Poland, extent of product innovation 33, 34*t*, 35*t*, 37*t*, 38
policy dimension 194–219
  financial constraints 211–12
  foreign direct investment 215–17
  grants 207–8
  infrastructure 213–14
  intellectual property rights and protection 196–203
  knowledge, access to 213
  prizes 218–19
  public contracts and government acquisition 208–11
  regulation 214–15
  subsidies 217–18
  tax incentives 203–7
political stability 79
Porter, M. E. 161
portfolio management 160
portfolios *see* product portfolios
Portugal, extent of product innovation 33, 34*t*, 35*t*, 37*t*
pre-emption effect 65, 67
Prennushi, G. 159
price indexes 168–79
  hedonic methods 174–8
pricing 73, 87
  demand for new products
    consumer durables 67
    consumer non-durables 68
    new-to-firm innovations 71
    new-to-market innovations 69–70, 71
    producer durables 66, 67

industry maturity 57, 58
interdependencies 73, 77–8
measurement 168–79
hedonic methods 174–8
price indexes 170–4
product creation and destruction patterns 168–70
new-to-market innovations 80
monopolies 80–3
oligopolies 84–5
original products 74–5
prizes 218–19
process innovations 1, 2
demand for a new product 63
design activities 20
determinants of 114
extent of 93, 108, 109, 110, 113, 115
firm performance 144, 150, 151–2
complementarities 157, 158
exports 153, 154–5
industry life cycle 160
productivity and R&D 150, 151–2
industry maturity 57, 58
knowledge flows 23
R&D 18
spending 91
surveys 33, 36, 43
welfare 184, 193
producer price indexes 173
producer products 125–6
defined 9
durables
demand for new products 63–7
diffusion of product innovations 126, 127
tax incentives 207
producers, defined 9
producer surplus 203 n. 2
intellectual property rights 200, 201, 203
welfare 180–91, 181*f*
product creation and destruction, patterns of 168–70
product development process 163–5
product differentiation 10
branding 10–11
entry to industries 121
firm performance 161–3
see also horizontal product differentiation; vertical product differentiation
product innovation, defined 11–14
productivity, estimates of impact of product innovation on 149–52, 153, 155
product launch data 44–54
book publishing 51–3
film industry 48–51
food industry 47–8
laundry detergents 46
music 53–4
pharmaceuticals 44–6
product life cycle (PLC) 62
diffusion of product innovations 124
firm performance 159–60
product portfolios 159–61
products, defined 8–11
product variety 191–3
profitability 136
accounting-based measures 139, 140
estimates of impact of product innovation on 142–6
innovation complementarities 157, 158
product differentiation 162
product life cycle 160
product portfolios 161
promotion 73, 87
interdependencies 73, 77–8
new-to-market innovations 80, 83, 86
original products 76
uncertainty 101
Prosser, L. 8, 8 n. 1
protective innovation 89, 90
public contracts 208–11
public sector borrowing requirement (PSBR) 206
publisher's right 29, 198
publishing 51–3, 52*t*
Pulkki-Brännström, A. M. 64, 128, 130

radical innovation 159
Ramey, G. 101, 192
Rammer, C. 157
rank effects 66, 67
Rao, H. 105
raw materials 9
real options theory 103
recorded music 53–4
Recording Industry of Japan 54
registered community design (RCD) 27, 198
regulations 214–15
as constraint on innovation 104, 117, 118*t*, 118, 119
entry and exit 121
location of production capacity 79
Reinganum, J. F. 67
Reisinger, M. 193
Reitsperger, W. D. 162

repeat purchase products *see* non-durable products
research and development (R&D) 16, 17–19
  applied research 16–17
    determinants of innovativeness 114
    incentives to product innovation 91
    spending on 18
  basic research 16, 17, 90–1
    determinants of innovativeness 114
    spending on 18
  Catapult centres 213
  costs 88
  data 16–17
  defined 17, 206
  demand for new products
    consumer durables 68
    producer durables 64
  and design activities, correlation between 20
  extent of product innovation 32, 111, 112, 113, 114–15, 116
  firm performance 146–7, 150–2, 164
  grant-based policies 207–8
  industry life cycle 120
  intellectual property rights 202
  intensity 121
  knowledge flows 22
  measurement 17
  public contracts 209
  risk diversification 102
  spending 17, 18
    extent of product innovation 115
    financial constraints 106–7
    intellectual property applications 29–30, 30*t*
    limitations as measure of innovation 91–2, 194
  subsidies 207
  tax incentives 115, 203–6
  welfare 192
residual income 139
resource-based view of the firm 164
Restieaux, A. 175
retailers, defined 9
Retail Price Index (RPI) 177
return on assets (ROA) 139, 143, 145
return on capital (ROC) 139
return on equity (ROE) 139
return on innovation investment (ROI2) 140–1
return on investment (ROI) 162
return on sales (ROS) 139, 144
Richter, A. 156
Rigobon, R. 195
risk
  as constraint on innovation 104, 105–7, 117, 118, 119
  launch aid 212
  policies to reduce 211–12
  in product innovation 101–3
  public contracts 209
Roberts, J. 156, 157
Roberts, P. W. 137, 142–3, 144
Robson, S. 20
Rodrik, D. 195
Rogers, E. M. 124, 129
Roller, L.-Hendrik 157
Romania, extent of product innovation 33, 34*t*, 35*t*, 37*t*
Rombach, D. 213
Romeo, A. A. 132
Rönn, M. 219
Roper, S. 20, 22, 110, 153
Rosenbaum, P. R. 154
Rosenberg, N. 130
Ross, S. A. 139
Rouvinen, P. 152
Rubin, D. B. 154
Rumelt, R. P. 144, 145
Russian Federation
  book publishing 52*t*
  extent of product innovation 43, 44*t*, 52*t*

Saggi, K. 93
sales margins 145–6
Salop, S. 96
Samsung 101
scale economies 192
scanner data
  extent of product innovation 54–6
  price indexes 174
  product creation and destruction 169
Scherer, F. M. 150
Schmalensee, R. 144, 145
Schmidt, T. 157
Schnaars, S. P. 25, 65
Schubert, T. 157
Schumpeter, J. A. 113, 136, 159
Schumpeterian hypotheses 113
Schwartz, S. 139
science
  policy 195
  and technology, interaction between 16
secrecy 200
Seed Enterprise Investment Scheme (SEIS) 211–12

Segway 214–15
Seim, K. 99
self-propagation of knowledge
  of new-to-market products 94
  of original products
    demand 63–4, 65, 69
    pricing 75
    promotion 76
Serbia, extent of product innovation 33, 34*t*, 35*t*, 37*t*
services
  diffusion of 131
  extent of product innovation 33–5, 35*t*
  and goods, distinction between 8
  R&D 18
Shaked, A. 10, 69, 82, 99, 162
Shaw, K. 159
Shelton, R. 137
Shuen, A. 164
Singhal, V. R. 149
Sinkula, J. M. 165
Skuras, D. 77 n. 1
Slater, S. F. 142, 163, 164, 165
Slovakia, extent of product innovation 34*t*, 35*t*, 37*t*
Slovenia
  extent of product innovation 34*t*, 35*t*, 37*t*
  firm performance 155
Smirlock, M. 139
software, sourcing new 111, 112
solar panel subsidies 217–18
Sorensen, A. T. 52
sources of new products 16–31
  creativity 20–1
  design activities 19–20
  emulation 24–30
  extent of product innovation 111–12
  imports and foreign direct investment 24
  knowledge flows 21–4
  R&D 17–19
South Africa
  extent of product innovation 44*t*
  Support Programme for Industrial Innovation (SPII) 207
South Korea *see* Korea
Spain
  extent of product innovation 33, 34*t*, 35*t*, 36*t*, 36, 37*t*
  firm performance 151, 153, 155
  patent box 205
spillover effects, as determinant of innovativeness 115
standard industrial classification (SIC) 8–9
standard reference index (SRI) 174
standards regulations 215
start-up firms 39, 119, 120–1
steel, diffusion of production of 133
Steffens, P. R. 132
Steiber, A. 129
Sterlacchini, A. 153
Stigler Commission 171 n. 1
Stiglitz, J. E. 84, 97, 106, 190, 192, 202
stock effects 65, 67
Stoneman, P. v, 2, 13, 14, 16, 32, 33, 53, 54, 62, 63, 64, 65, 72, 74, 75, 82, 105, 109, 110, 111, 115, 116, 124, 128, 129, 130, 131, 132, 158, 185, 188–9, 190, 202, 206, 212
strategic innovation
  complementarities 158
  extent of 109, 110
Stumpfe, J. 58
subsidies 91, 207–8, 217–18
  as determinant of innovativeness 115
  launch aid 212
Sullivan, M. W. 162
Sundaram, A. K. 149
suppliers, defined 9
surveys of product innovation 32–44
  BRIC and developing countries 43–4
  Europe 32–41
  US 41–3
Sutton, J. 10, 58, 69, 82, 83, 85, 99, 100, 101, 162
Svenson, R. A. 137
Sweden
  extent of product innovation 33, 34*t*, 35*t*, 37*t*
  firm performance 145–6
  price indexes 177

Taiwan
  Apple 78
  diffusion of production of product innovations 133
Takalo, T. 207, 208
tariffs 79, 196
Tarola, O. 90
tax incentives
  as determinant of innovativeness 115
  foreign direct investment 216–17
  policy dimension 203–7, 212, 216–17
  venture capital 212
tax regimes 79
technological innovation
  complementarities 157, 158
  extent of 110–11

technology
as determinant of innovativeness 114, 115
diffusion 62
adoption/adoption lag 121–2, 127–8
extensive margin 126–8, 129
intensive margin 127, 128–30
intra-firm 129, 131–2
intra-household 129, 131–2
pull/push factors 89, 91
and science, interaction between 16
Teece, D. J. 164
television sets, diffusion of production of 132–3
Tellis, G. J. 124
Tether, B. S. 20
Thailand
book publishing 52*t*
extent of product innovation 52*t*
Thisse, J.-F. 10, 69, 162
Tidd, J. 24
Tilton, J. E. 120
timing of product innovation 103
Tirole, J. 65
Tisdell, C. 132
Tobin's Q 140
Toivanen, O. 147
Topkis, D. M. 156
total market value to capital employed (TMV/CE) ratio 145
trademarks 26, 29
extent of product innovation 47–8, 48*t*
Madrid system *see* Madrid trademark system
policy dimension 197, 198–9, 200
R&D spending 29–30, 30*t*
registered/unregistered 29, 198
registrations and renewals (1970–2009) 48, 49*t*
trade policy 196
Trajtenberg, M. 14, 184
transport costs 79
Triplett, J. E. 172, 173
Tsano, G. 164
Turkey, extent of product innovation 33, 34*t*, 35*t*, 37*t*
turnover, contribution of product innovation to 39–41, 40–1*t*

UK Trade and Investment 216
Ulrich, K. T. 137
uncertainty in product innovation 101–3, 105
dominant format policies 215
industry life cycle 120
public contracts 209
UNESCO Institute for Statistics (UIS) 50, 51
United Kingdom
Arts Council 208
Association of the British Pharmaceutical Industry 211
book publishing 52, 52*t*
Catapult centres 213
constraints on innovation 105
copyright 28–9, 198, 201
Department of Health (DH) 211
Department of Trade and Industry 213
design rights 27, 197–8
diffusion of product innovations 128, 132–3, 134
Enterprise Investment Scheme (EIS) 211–12
extent of product innovation 109–11
determinants 116
product launch data 46, 48, 52, 52*t*, 53, 54
surveys 33, 34*t*, 37*t*, 38–41, 38*t*, 40–1*t*, 43
financial constraints, policies to address 211–12
firm performance
exports 153–4
innovation complementarities 158
manufacturing sector 143–4
market value 146, 147
productivity 152
food industry 48
foreign direct investment 216
grant-based policies 207, 208
infrastructure
electric car charging stations 214
mobile telephony 213
Innovate UK 213
Innovation Surveys
capacity creation 76–7
constraints on innovation 117–18, 118*t*
costs of getting new products to market 88
design activities 19–20
diffusion of product innovations 131
extent of product innovation 33, 34*t*, 37*t*, 38–41, 38*t*, 40–1*t*, 43, 109
financial constraints to innovation 105–6
innovation complementarities 158
intellectual property rights 199–200
knowledge flows 22–3, 23*t*
Intellectual Property Office 28, 198–9, 201, 205
Trade Marks Register 29, 198–9

internal finance as determinant of firm's innovativeness 114
knowledge, policies to facilitate access to 213
launch aid 212
laundry detergents 46
music 53, 54
National Health Service (NHS) 211
National Institute of Health and Care Excellence (NICE) 211
Office of National Statistics (ONS) 170
patents 197
Pharmaceutical Price Regulation Scheme (PPRS) 211
policy dimension 194
price indexes 170, 177
prizes 218–19
public contracts 210–11
regulations 214–15
Seed Enterprise Investment Scheme (SEIS) 211–12
standard industrial classification (SIC) 8–9
subsidies 217–18
tax incentives 204, 206
levy grant system 206
patent box 205–6
trademarks 29
United Nations Conference on Trade and Development (UNCTAD) 216
United States
aerospace industry 212
Apple 78
book publishing 52, 52*t*
Bureau of Labor Statistics (BLS) 170, 172, 176
copyright 29, 198
design patents 26, 197
diffusion of production of product innovations 132–3, 134
exhaust emission regulations 214
extent of product innovation
product launch data 44–6, 45*t*, 46*t*, 47, 48, 50, 50*t*, 51, 52, 52*t*, 53
scanner data 55–6
surveys 41–3, 42*t*
Federal Communications Commission (FCC) 213
film industry 50, 50*t*, 51
firm performance
market value 146, 147, 148, 149
pharmaceutical sector 143
Food and Drug Administration (FDA) 44–6, 46*t*
food industry 47, 48
industry development 58–9
internal finance as determinant of firm's innovativeness 114
music 53
pharmaceuticals 44–6, 45*t*, 46*t*
price indexes 170, 171 n. 1, 178, 179
Boskin Commission 171–2
hedonic methods 176, 177
Stigler Commission 171 n. 1
product creation and destruction 169–70
Small Business Innovation Research (SBIR) program 114
tax incentives 206
wage levels 79
unregistered community design 27, 198
Uruguay, extent of product innovation 44*t*
utility patents 26
Utterback, J. M. 57, 58
Uyarra, E. 208

Valentini, G. 22
Van Ark, B. 177 n. 3
Van der Werf, P. A. 25, 65
Van Reenen, J. 126
Veblen, T. 67
vehicles, diffusion of production of 133–4
Venkatraman, N. 142
venture capital 211–12
Vermeulen, P. A. M. 113
Vernon, R. 152
vertical product differentiation 10
branding 11
defined 14
demand for new products 69–70
extent of 99–101
firm performance 161–3
new-to-market innovations 82–3, 85–6
welfare 182–3, 192–3
video games 54
Vinding, A. L. 1

wage levels 79
Wakelin, K. 153
Wall, G. 47
Warzynski, F. 151
Webster, F. E., Jr 163
Weinstein, D. E. 54–5, 56, 60, 169–70, 172, 174, 178
Weiss, A. 106
welfare 180–93
consumer and producer surplus 180–91
intellectual property rights 202

welfare (*cont.*)
  product variety 191–3
  subsidies 208
Wells, J. 175
Wernerfelt, B. 164
Whittington, R. 159
wholesalers, defined 9
Wiersema, F. D. 161
Williams, J. 202
Winger, R. 47
Woodruff, R. B. 162
World Intellectual Property Organisation (WIPO)
  design rights 27
  food industry 47, 48
  intellectual property defined 196
  trademarks 199
World Trade Organization (WTO) 212

Xuereb, J.-M. 164, 165

Zhu, D. H. 162